AWWA Wastewater Operator Field Guide

Compiled by AWWA staff members:
John M. Stubbart
William C. Lauer
Timothy J. McCandless
Paul Olson

American Water Works
Association

Science and Techn

Project Manager: Melissa Christensen, Senior Technical Editor
Produced by Glacier Publishing Services, Inc.

Disclaimer
The authors, contributors, editors, and publisher do not assume responsibility for the validity of the content or any consequences of their use. In no event will AWWA be liable for direct, indirect, special, incidental, or consequential damages arising out of the use of information presented in this book. In particular, AWWA will not be responsible for any costs, including, but not limited to, those incurred as a result of lost revenue. In no event shall AWWA's liability exceed the amount paid for the purchase of this book.

**Library of Congress Cataloging-in-Publication Data
has been applied for.**

ISBN 1-58321-386-4

**American Water Works
Association**

6666 West Quincy Avenue
Denver, CO 80235-3098
303.794.7711

Contents

Preface

This guide is a compilation of information, charts, graphs, formulas, and definitions that are used by wastewater system operators in performing their daily duties. There is so much information contained in so many different sources that finding it while in the field can be a problem. This guide compiles information mostly from AWWA manuals, books, and standards, but also from other generic information found in many publications.

The sections of this guide group the information based on how it would be used by the operator. The guide includes information for both wastewater treatment and collection. Design engineers should also find this material helpful. Major sections include math, conversion factors, chemistry, safety, collection, pumps and motors, flow, wastewater treatment, biosolids, and disposal. Perusing the guide now will assist in finding handy information later. This is the first edition of the guide. If you would like to suggest changes or additions to the guide, please submit them to AWWA, Publishing Group, 6666 W. Quincy Ave., Denver, CO 80235.

Basic Math

A number of calculations are used in the operation of small wastewater facilities. Some only need to be calculated once and recorded for future reference; others may need to be calculated more frequently. Operators need to be familiar with the formulas and basic calculations to carry out their duties properly. Note that the formulas in this section are basic and general; specific formulas for particular components of wastewater systems can be found in the relevant sections of this guide.

SYSTÈME INTERNATIONAL UNITS _____

When performing calculations, water operators should pay particular attention not only to the numbers but also to the units involved. Where SI units and customary units are given, convert all units to one system, usually SI, *first*. Be sure to write the appropriate units with each number in the calculations for clarity. Inaccurate calculations and measurements can lead to incorrect reports and costly operational decisions. This section introduces the calculations that are the basic building blocks of the water/wastewater industry.

SI Prefixes

The SI is based on factors of ten, similar to the dollar. This allows the size of the unit of measurement to be increased or decreased while the base unit remains the same. The SI prefixes are

$$\begin{aligned}
\text{mega, M} &= 1,000,000 \times \text{the base unit} \\
\text{kilo, k} &= 1,000 \times \text{the base unit} \\
\text{hecta, h} &= 100 \times \text{the base unit} \\
\text{deca, da} &= 10 \times \text{the base unit} \\
\text{deci, d} &= 0.1 \times \text{the base unit} \\
\text{centi, c} &= 0.01 \times \text{the base unit} \\
\text{milli, m} &= 0.001 \times \text{the base unit} \\
\text{micro, } \mu &= 0.000001 \times \text{the base unit}
\end{aligned}$$

Base SI Units

Quantity	Unit	Abbreviation
length	meter	m
mass	kilogram	kg
time	second	sec
electric current	ampere	A
thermodynamic temperature	kelvin	K
amount of substance	mole	mol
luminous intensity	candela	cd

Supplementary SI Units

Quantity	Unit	Abbreviation
plane angle	radian	rad
solid angle	steradian	sr

Derived SI Units With Special Names

Quantity	Unit	Abbreviation	Equivalent-Units Abbreviation
frequency (of a periodic phenomenon)	hertz	Hz	sec^{-1}
force	newton	N	$kg \cdot m/sec^2$
pressure, stress	pascal	Pa	N/m^2
energy, work, quantity of heat	joule	J	$N \cdot m$
power, radiant flux	watt	W	J/sec
quantity of electricity, electric charge	coulomb	C	$A \cdot sec$
electric potential, potential difference, electromotive force	volt	V	W/A
electrical capacitance	farad	F	C/V
electrical resistance	ohm	Ω	V/A
electrical conductance	siemens	S	A/V
magnetic flux	weber	Wb	$V \cdot sec$
magnetic flux density	tesla	T	Wb/m^2
inductance	henry	H	Wb/A
luminous flux	lumen	lm	$cd \cdot Sr$
luminance	lux	lx	lm/m^2
activity (of a radionuclide)	becquerel	Bq	disintegrations/sec
absorbed ionizing radiation dose	gray	Gy	J/kg
ionizing radiation dose equivalent	sievert	Sv	J/kg

Some Common Derived SI Units

Quantity	Unit	Abbreviation
absorbed dose rate	grays per second	Gy/sec
acceleration	meters per second squared	m/sec^2
angular acceleration	radians per second squared	rad/sec^2
angular velocity	radians per second	rad/sec
area	square meter	m^2
concentration (amount of substance)	moles per cubic meter	mol/m^3
current density	amperes per square meter	A/m^2
density, mass	kilograms per cubic meter	kg/m^3
electric charge density	coulombs per cubic meter	C/m^3
electric field strength	volts per meter	V/m
electric flux density	coulombs per square meter	C/m^2
energy density	joules per cubic meter	J/m^3
entropy	joules per kelvin	J/K
exposure (X and gamma rays)	coulombs per kilogram	C/kg
heat capacity	joules per kelvin	J/K
heat flux density irradiance	watts per square meter	W/m^2
luminance	candelas per square meter	cd/m^2
magnetic field strength	amperes per meter	A/m
molar energy	joules per mole	J/mol
molar entropy	joules per mole kelvin	J/(mol·K)
molar heat capacity	joules per mole kelvin	J/(mol·K)
moment of force	newton-meter	N·m
permeability (magnetic)	henrys per meter	H/m
permittivity	farads per meter	F/m
power density	watts per square meter	W/m^2

Table continued on next page

4

Some Common Derived SI Units (continued)

Quantity	Unit	Abbreviation
radiance	watts per square meter steradian	$W/(m^2 \cdot sr)$
radiant intensity	watts per steradian	W/sr
specific energy	joules per kilogram	J/kg
specific entropy	joules per kilogram kelvin	$J/(kg \cdot K)$
specific heat capacity	joules per kilogram kelvin	$J/(kg \cdot K)$
specific volume	cubic meters per kilogram	m^3/kg
surface tension	newtons per meter	N/m
thermal conductivity	watts per meter kelvin	$W/(m \cdot K)$
velocity	meters per second	m/sec
viscosity, absolute	pascal-second	$Pa \cdot sec$
viscosity, kinematic	square meters per second	m^2/sec
volume	cubic meter	m^3
wave number	per meter	m^{-1}

KEY FORMULAS FOR MATH

Area Formulas

Square

area = s × s
diagonal = 1.414 × s

Rectangle or Parallelogram

area = b × h
diagonal = square root $(b^2 + h^2)$

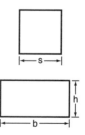

Trapezoid

$$\text{area} = \frac{(a + b)h}{2}$$

Any Triangle

$$\text{area} = \frac{b \times h}{2}$$

Right-Angle Triangle

$$a^2 + b^2 = c^2$$

Circle

area = $\pi \times r^2$
circumference = $2 \times \pi \times r$

Sector of a Circle

$$\text{area} = \frac{\pi \times r \times r \times \alpha}{360}$$

$$\text{length} = 0.01745 \times r \times \alpha$$

$$\text{angle} = \frac{1}{0.01745 \times r}$$

$$\text{radius} = \frac{1}{0.01745 \times \alpha}$$

Ellipse

area = $\pi \times a \times b$

Volume Formulas

$$\text{rectangle tank volume} = \left(\begin{array}{c}\text{area of}\\\text{rectangle}\end{array}\right)\left(\begin{array}{c}\text{third}\\\text{dimension}\end{array}\right)$$

$$= \text{lw}\left(\begin{array}{c}\text{area of}\\\text{rectangle}\end{array}\right)$$

$$\text{trough volume} = \left(\begin{array}{c}\text{area of}\\\text{rectangle}\end{array}\right)\left(\begin{array}{c}\text{third}\\\text{dimension}\end{array}\right)$$

$$= \left(\frac{\text{bh}}{2}\right)\left(\begin{array}{c}\text{third}\\\text{dimension}\end{array}\right)$$

$$\text{cylinder volume} = \left(\begin{array}{c}\text{area of}\\\text{rectangle}\end{array}\right)\left(\begin{array}{c}\text{third}\\\text{dimension}\end{array}\right)$$

$$= (0.785\ \text{D}^2)\left(\begin{array}{c}\text{third}\\\text{dimension}\end{array}\right)$$

$$\text{cone volume} = \frac{1}{3}\ (\text{volume of a cylinder})$$

$$\text{sphere volume} = \left(\frac{\pi}{6}\right)(\text{diameter})^3$$

Rectangular Solid

volume = $h \times a \times b$
surface area = $(2 \times a \times b) + (2 \times b \times h) +$
$\qquad\qquad\quad (2 \times a \times b)$

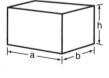

Cylinder

volume = $\pi \times r^2 \times h$
surface area = $2 \times \pi \times rh$
$\pi = 3.142$

Elliptical Cylinder

volume $= \pi \times a \times b \times h$

area $= 6.283 \times \dfrac{\sqrt{a^2 + b^2}}{2} \times h + 6.283 \times a \times b$

Sphere

volume $= \dfrac{4 \times \pi \times r^3}{3}$

surface area $= 4 \times \pi \times r^2$

Cone

volume $= \dfrac{\pi \times r^2 \times h}{3}$

surface area $= \pi \times r \times \sqrt{r^2} \times (r + h) \times h$

Pyramid

volume $= \dfrac{a \times b \times h}{3}$

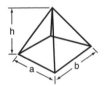

Other Formulas

$$\frac{\text{theoretical water}}{\text{horsepower}} = \frac{\text{gal/min} \times \text{total head, ft}}{3{,}960}$$

$$= \frac{\text{gal/min} \times \text{lb/in.}^2}{1{,}715}$$

$$\text{brake horsepower} = \frac{\text{theoretical water horsepower}}{\text{pump efficiency}}$$

$$\text{detention time, min} = \frac{\text{volume of basin, gal}}{\text{flow rate, gpm}}$$

$$\text{filter backwash rate, gal/min/ft}^2 = \frac{\text{flow, gpm}}{\text{area of filter, ft}^2}$$

$$\text{surface overflow rate} = \frac{\text{flow, gpm}}{\text{area, ft}^2}$$

$$\text{weir overflow rate} = \frac{\text{flow, gpm}}{\text{weir length, ft}}$$

$$\text{pounds per mil gal} = \text{parts per million} \times 8.34$$

$$\text{parts per million} = \text{pounds per mil gal} \times 0.12$$

$$\text{parts per million} = \text{percent strength of solution} \times 10,000$$

$$\text{pounds per day} = \text{volume, mgd} \times \text{dosage, mg/L} \times 8.34 \text{ lb/gal}$$

$$\text{dosage, mg/L} = \frac{\text{feed, lb/day}}{\text{volume, mgd} \times 8.34 \text{ lb/gal}}$$

$$\text{percent element by weight} = \frac{\text{weight of element in compound}}{\text{molecular weight of compound}} \times 100$$

$$\text{rectangular basin volume, ft}^3 = \text{length, ft} \times \text{width, ft} \times \text{height, ft}$$

$$\text{rectangular basin volume, gal} = \text{length, ft} \times \text{width, ft} \times \text{height, ft} \times 7.48 \text{ gal/ft}^3$$

$$\text{right cylinder volume, ft}^3 = 0.785 \times \text{diameter}^2\text{, ft} \times \text{height or depth, ft}$$

$$\text{right cylinder volume, gal} = 0.785 \times \text{diameter}^2\text{, ft} \times \text{height or depth, ft} \times 7.48 \text{ gal/ft}^3$$

$$\text{gallons per capita per day, average water usage} = \frac{\text{volume, gpd}}{\text{population served/day}}$$

$$\frac{\text{supply, days}}{\text{(full to tank dry)}} = \frac{\text{volume, gpd}}{\text{population served} \times \text{gpcd}}$$

$$\begin{array}{c}\text{gallons per day of} \\ \text{water consumption,} \\ \text{(demand/day)}\end{array} = \text{population} \times \text{gpcd}$$

Consumption Averages, per capita

winter = 170 gpcd
spring = 225 gpcd
summer = 325 gpcd

KEY CONVERSIONS FOR FLOWS

Conversion of US customary flow units can be easily made using the block diagram below. When moving from a smaller to a larger block, multiply by the factor shown on the connecting line. When moving from a larger to a smaller block, divide.

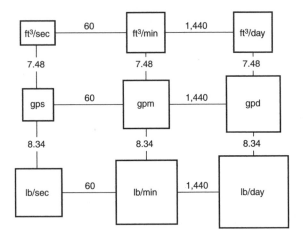

$$\text{flow, gpm} = \text{flow, cfs} \times 448.8 \text{ gpm/cfs}$$

$$\text{flow, cfs} = \frac{\text{flow, gpm}}{448.8 \text{ gpm/cfs}}$$

pipe diameter, in. $= \dfrac{\text{area, ft}^2}{0.785} \times 12 \text{ in./ft}$

actual leakage, gpd/mi./in. $= \dfrac{\text{leak rate, gpd}}{\text{length, mi.} \times \text{diameter, in.}}$

NOTE: minimum flushing velocity: 2.5 fps

maximum pipe velocity: 5.0 fps

key conversions: 1.55 cfs/mgd; 448.8 gpm/cfs

KEY FORMULAS FOR FLOWS AND METERS

Velocity

flow, cfs $=$ area, ft $\times$ velocity, fps

$\dfrac{\text{gpm}}{448.8 \text{ gpm/cfs}} = 0.785 \times \text{diameter, ft}^2 \times \dfrac{\text{distance, ft}}{\text{time, sec}}$

velocity, fps $= \dfrac{\text{flow, cfs}}{\text{area, ft}^2}$

area, ft^2 $= \dfrac{\text{flow, cfs}}{\text{velocity, fps}}$

Head Loss Resulting From Friction

Darcy–Weisbach Formula

$$h_L = f(L/D)(V^2/2g)$$

Where (in any consistent set of units):

h_L = head loss

f = friction factor, dimensionless

L = length of pipe

D = diameter of the pipe

V = average velocity

g = gravity constant

Flow Rate Calculations

The *rule of continuity* states that the flow Q that enters a system must also be the flow that leaves the system.

$$Q_1 = Q_2 \quad \text{or} \quad A_1V_1 = A_2V_2$$

or

$$Q = AV$$

Where:

$\quad Q$ = flow rate

$\quad A$ = area

$\quad V$ = velocity

$$\begin{pmatrix} \text{flow rate,} \\ \text{ft}^3/\text{time} \end{pmatrix} = \begin{pmatrix} \text{width,} \\ \text{ft} \end{pmatrix} \times \begin{pmatrix} \text{depth,} \\ \text{ft} \end{pmatrix} \times \begin{pmatrix} \text{velocity,} \\ \text{ft/time} \end{pmatrix}$$

$$\begin{pmatrix} \text{feed rate,} \\ \text{lb/day} \end{pmatrix} = \begin{pmatrix} \text{dosage,} \\ \text{ppm} \end{pmatrix} \times \begin{pmatrix} \text{flow rate,} \\ \text{mgd} \end{pmatrix} \times \begin{pmatrix} \text{conversion factor} \\ \text{8.34 lb/gal} \end{pmatrix}$$

$$\begin{pmatrix} \text{chlorine} \\ \text{weight, lb} \end{pmatrix} = \begin{pmatrix} \text{dosage,} \\ \text{mg/L} \end{pmatrix} \times \begin{pmatrix} \text{volume of} \\ \text{container} \\ \text{mil/gal} \end{pmatrix} \times \begin{pmatrix} \text{conversion factor} \\ \text{8.34 lb/gal} \end{pmatrix}$$

Summary of Pressure Requirements

Requirement	Value		Location
	psi	*(kPa)*	
Minimum pressure	35	(241)	All points within distribution system
	20	(140)	All ground level points
Desired maximum	100	(690)	All points within distribution system
Fire flow minimum	20	(140)	All points within distribution system
Ideal range	50–75	(345–417)	Residences
	35–60	(241–414)	All points within distribution system

Units of Measure and Conversions

The ability to accurately and consistently measure such variables as flow and head, along with wastewater quality indicators such as chemical and biological oxygen demand, total suspended solids, toxins, and pathogens is a key component of the successful operation of a wastewater distribution system. This section provides the most common units of measure and associated conversions typically used in the wastewater industry.

UNITS OF MEASURE

acre An SI unit of area.

acre-foot (acre-ft) A unit of volume. One acre-foot is the equivalent amount or volume of water covering an area of 1 acre that is 1 foot deep.

ampere (A) An SI unit of constant current that, if maintained in two straight parallel conductors of infinite length or negligible cross section and placed 1 meter apart in a vacuum, would produce a force equal to 2×10^{-7} newtons per meter of length.

ampere-hour (A·hr) A unit of electric charge equal to 1 ampere flowing for 1 hour.

angstrom (Å) A unit of length equal to 10^{-10} meter.

atmosphere (atm) A unit of pressure equal to 14.7 pounds per square inch (101.3 kilopascals) at average sea level under standard conditions.

bar A unit of pressure defined as 100 kilopascals.

barrel (bbl) A unit of volume, frequently 42 gallons for petroleum or 55 gallons for water.

baud A measure of analog data transmission speed that describes the modulation rate of a wave, or the average frequency of the signal. One baud equals 1 signal unit per second. If an analog signal is viewed as an electromagnetic wave, one complete wavelength or cycle is equivalent to a signal unit. The term *baud* has often been used synonymously with *bits per second*. The baud rate may equal bits per second for some transmission techniques, but special modulation techniques frequently deliver a bits-per-second rate higher than the baud rate.

becquerel (Bq) An SI unit of the activity of a radionuclide decaying at the rate of one spontaneous nuclear transition per second.

billion electron volts (BeV) A unit of energy equivalent to 10^{9} electron volts.

billion gallons per day (bgd) A unit for expressing the volumetric flow rate of water being pumped, distributed, or used.

binary digits (bits) per second (bps) A measure of the data transmission rate. A binary digit is the smallest unit of information or data, represented by a binary "1" or "0."

British thermal unit (Btu) A unit of energy. One British thermal unit was formerly defined as the quantity of heat required to raise the temperature of 1 pound of pure water 1° Fahrenheit; now defined as 1,055.06 joules.

bushel (bu) A unit of volume.

caliber (1) The diameter of a round body, especially the internal diameter of a hollow cylinder. (2) The diameter of a bullet or other projectile, or the diameter of a gun's bore. In US customary units, usually expressed in hundredths or thousandths of an inch and typically written as a decimal fraction (e.g., 0.32). In SI units, expressed in millimeters.

calorie (gram calorie) A unit of energy. One calorie is the amount of heat necessary to raise the temperature of 1 gram of pure water at 15° Celsius by 1° Celsius.

candela (cd) An SI unit of luminous intensity. One candela is the luminous intensity, in a given direction, of a source that emits monochromatic radiation of frequency 540×10^{12} hertz and that has a radiant intensity in that direction of $1/683$ watt per steradian.

candle A unit of light intensity. One candle is equal to 1 candela. Candelas are the preferred units.

candlepower A unit of light intensity. One candlepower is equal to 1 candela. Candelas are the preferred units.

centimeter (cm) A unit of length defined as one hundredth of a meter.

centipoise A unit of absolute viscosity equivalent to 10^{-2} poise. See also *poise*.

chloroplatinate (Co–Pt) unit (cpu) See *color unit.*

cobalt–platinum unit See *color unit.*

colony-forming unit (cfu) A unit of expression used in enumerating bacteria by plate-counting methods. A colony of bacteria develops from a single cell or a group of cells, either of which is a colony-forming unit.

color unit (cu) The unit used to report the color of water. Standard solutions of color are prepared from potassium chloroplatinate (K_2PtCl_6) and cobaltous chloride ($CoCl_2 \cdot 6H_2O$). Adding the following amounts in 1,000 milliliters of distilled water produces a solution with a color of 500 color units: 1.246 grams potassium chloroplatinate, 1.00 grams geobaltous chloride, and 100 milliliters concentrated hydrochloric acid (HCl).

coulomb (C) An SI unit of a quantity of electricity or electric charge. One coulomb is the quantity of electricity transported in 1 second by a current of 1 ampere, or about 6.25×10^{18} electrons. Coulombs are equivalent to ampere-seconds.

coulombs per kilogram (C/kg) A unit of exposure dose of ionizing radiation. See also *roentgen.*

cubic feet (ft^3) A unit of volume equivalent to a cube with a dimension of 1 foot on each side.

cubic feet per hour (ft^3/hr) A unit for indicating the rate of liquid flow past a given point.

cubic feet per minute (ft³/min, CFM) A unit for indicating the rate of liquid flow past a given point.

cubic feet per second (ft³/sec, cfs) A unit for indicating the rate of liquid flow past a given point.

cubic inch (in.³) A unit of volume equivalent to a cube with a dimension of 1 inch on each side.

cubic meter (m³) A unit of volume equivalent to a cube with a dimension of 1 meter on each side.

cubic yard (yd³) A unit of volume equivalent to a cube with a dimension of 1 yard on each side.

curie (Ci) A unit of radioactivity. One curie equals 37 billion disintegrations per second, or approximately the radioactivity of 1 gram of radium.

cycles per second (cps) A unit for expressing the number of times something fluctuates, vibrates, or oscillates each second. These units have been replaced by hertz. One hertz equals 1 cycle per second.

dalton (D) A unit of weight. One dalton designates $1/16$ the weight of oxygen-16. One dalton is equivalent to 0.9997 atomic weight unit, or nominally 1 atomic weight unit.

darcy (da) The unit used to describe the permeability of a porous medium (e.g., the movement of fluids through underground formations studied by petroleum engineers, geologists or geophysicists, and groundwater specialists). A porous medium is said to have a permeability of 1 darcy if a fluid of 1-centipoise viscosity that completely fills the pore space of the medium will flow through it at a rate of 1 cubic centimeter per second per square centimeter of cross-sectional area under a pressure gradient of 1 atmosphere per centimeter of length. In SI units, 1 darcy = 9.87×10^{-13} square meters.

day A unit of time equal to 24 hours.

decibel (dB) A dimensionless ratio of two values expressed in the same units of measure. It is most often applied to a power ratio and defined as decibels = 10 $\log_{10}$ (actual power level/reference power level), or dB = 10 $\log_{10}$ (W_2/W_1), where W is the power level in watts per square centimeter for sound. Power is proportional to the square of potential. In the case of sound, the potential is measured as a pressure, but the sound level is an energy level. Thus, dB = 10 $\log_{10}$ (p_2/p_1)^2 or dB = 20 $\log_{10}$ (p_2/p_1), where p is the potential. The reference levels are not well standardized. For example, sound power is usually measured above 10^{-12} watts per square centimeter, but both 10^{-11} and 10^{-16} watts per square centimeter are used. Sound pressure is usually measured above 20 micropascals in air. The reference level is not important in most cases because one is usually concerned with the difference in levels, i.e., with a power ratio. A power ratio of 1.26 produces a difference of 1 decibel.

deciliter (dL) A unit of volume defined as one tenth of a liter. This unit is often used to express concentration in clinical chemistry. For example, a concentration of lead in blood would typically be reported in units of micrograms per deciliter.

degree (°) A measure of the phase angle in a periodic electrical wave. One degree is $1/360$ of the complete cycle of the periodic wave. Three hundred sixty degrees equals 2π radians.

degree Celsius (°C) A unit of temperature. The degree Celsius is exactly equal to the kelvin and is used in place of the kelvin for expressing Celsius temperature (symbol t) defined by the equation $t = T - T_0$, where T is the thermodynamic temperature in kelvin and $T_0 = 273.15$ kelvin by definition.

degree Fahrenheit (°F) A unit of temperature on a scale in which 32° marks the freezing point and 212° the boiling point of water at a barometric pressure of 14.7 pounds per square inch.

degree kelvin (K) *See* kelvin.

dram (dr) Small weight. Two different drams exist: the apothecary's dram (equivalent to 1/3.54 gram) and the avoirdupois dram (equivalent to 1/1.17 gram).

electron volt (eV) A unit of energy commonly used in the fields of nuclear and high-energy physics. One electron volt is the energy transferred to a charged particle with a single charge when that particle falls through a potential of 1 volt. An electron volt is equal to 1.6×10^{-19} joule.

equivalents per liter (eq/L) An SI unit of an expression of concentration equivalent to normality. The normality of a solution (equivalent weights per liter) is a convenient way of expressing concentration in volumetric analyses.

fathom A unit of length equivalent to 6 feet, used primarily in marine measurements.

feet (ft) The plural form of a unit of length (the singular form is *foot*).

feet board measure (fbm) A unit of volume. One board foot is represented by a board measuring 1 foot long by 1 foot wide by 1 inch thick (144 cubic inches). A board measuring 0.5 feet by 2 feet by 2 inches thick would equal 2 board feet.

feet per hour (ft/hr) A unit for expressing the rate of movement.

feet per minute (ft/min) A unit for expressing the rate of movement.

feet per second (ft/sec, fps) A unit for expressing the rate of movement.

feet per second squared (ft/sec^2) A unit of acceleration (the rate of change of linear motion). For example, the acceleration caused by gravity is 32.2 ft/sec^2 at sea level.

feet squared per second (ft^2/sec) A unit used in flux calculations.

fluid ounce (fl oz) A unit for expressing volume, equivalent to $1/128$ of a gallon.

foot A unit of length, equivalent to 12 inches. See also *US customary system of units*.

foot of water (39.2° Fahrenheit) A unit for expressing pressure or elevation head.

foot per second per foot (ft/sec/ft; sec^{-1}) A unit for expressing velocity gradient.

foot-pound, torque A unit for expressing the energy used in imparting rotation, often associated with the power of engine-driven mechanisms.

foot-pound, work A unit of measure of the transference of energy when a force produces movement of an object.

formazin turbidity unit (ftu) A turbidity unit appropriate when a chemical solution of formazin is used as a standard to calibrate a turbidimeter. If a nephelometric turbidimeter is used, nephelometric turbidity units and formazin turbidity units are equivalent. See also *nephelometric turbidity unit*.

gallon (gal) A unit of volume, equivalent to 231 cubic inches. See also *Imperial gallon*.

gallons per capita per day (gpcd) A unit typically used to express the average number of gallons of water used by the average person each day in a water system. The calculation is made by dividing the total gallons of water used each day by the total number of people using the water system.

gallons per day (gpd) A unit for expressing the discharge or flow past a fixed point.

gallons per day per square foot (gpd/ft^2, gsfd) A unit of flux equal to the quantity of liquid in gallons per day through 1 square foot of area. It may also be expressed as a velocity in units of length per unit time. In pressure-driven membrane treatment processes, this unit is commonly used to describe the volumetric flow rate of permeate through a unit area of active membrane surface. In settling tanks, this rate is called the overflow rate.

gallons per flush (gal/flush) The number of gallons used with each flush of a toilet.

gallons per hour (gph) A unit for expressing the discharge or flow of a liquid past a fixed point.

gallons per minute (gpm) A unit for expressing the discharge or flow of a liquid past a fixed point.

gallons per minute per square foot (gpm/ft^2) A unit for expressing flux, the discharge or flow of a liquid through a unit of area. In a filtration process, this unit is commonly used to describe the volumetric flow rate of filtrate through a unit of filter media surface area. It may also be expressed as a velocity in units of length per unit time.

gallons per second (gps) A unit for expressing the discharge or flow past a fixed point.

gallons per square foot (gal/ft^2) A unit for expressing flux, the discharge or flow of a liquid through each unit of surface area of a granular filter during a filter run (between cleaning or backwashing).

gallons per square foot per day See *gallons per day per square foot.*

gallons per year (gpy) A unit for expressing the discharge or flow of a liquid past a fixed point.

gamma (γ) A symbol used to represent 1 microgram. Avoid using this symbol; the preferred symbol is µg.

gigabyte (GB) A unit of computer memory. One gigabyte equals 1 megabyte times 1 kilobyte, or 1,073,741,824 bytes (roughly 1 billion bytes).

gigaliter (GL) A unit of volume defined as 1 billion liters.

grad A unit of angular measure equal to $1/400$ of a circle.

grain (gr) A unit of weight.

grains per gallon (gpg) A unit sometimes used for reporting water analysis concentration results in the United States and Canada.

gram (g) A fractional unit of mass. One gram was originally defined as the weight of 1 cubic centimeter or 1 milliliter of water at 4° Celsius. Now it is $1/1,000$ of the mass of a certain block of platinum–iridium alloy known as the international prototype kilogram, preserved at Sèvres, France.

gram molecular weight The molecular weight of a compound in grams. For example, the gram molecular weight of CO_2 is 44.01 grams. See also *mole.*

gray (Gy) An SI unit of *absorbed* ionizing radiation dose. One gray, equal to 100 rad, is the absorbed dose when the energy per unit mass imparted to matter by ionizing radiation is 1 joule per kilogram. See also *rad*; *rem*; *sievert.*

hectare (ha) A unit of area equivalent to 10,000 square meters.

henry (H) An SI unit of electric inductance, equivalent to meters squared kilograms per second squared per ampere squared. One henry is the inductance of a closed circuit in which an electromotive force of 1 volt is produced when the electric current in the circuit varies uniformly at a rate of 1 ampere per second.

hertz (Hz) An SI unit of measure of the frequency of a periodic phenomenon in which the period is 1 second, equivalent to second^{-1}. Hertz units were formerly expressed as cycles per second.

horsepower (hp) A standard unit of power. See also *US customary system of units*.

horsepower-hour (hp·hr) A unit of energy or work.

hour (hr) An interval of time equal to $1/24$ of a day.

Imperial gallon A unit of volume used in the United Kingdom, equivalent to the volume of 10 pounds of freshwater.

inch (in.) A unit of length.

inch of mercury (32° Fahrenheit) A unit of pressure or elevation head.

inch-pound (in.-lb) A unit of energy or torque.

inches per minute (in./min) A unit of velocity.

inches per second (in./sec) A unit of velocity.

International System of Units. See *Système International*.

joule (J) An SI unit for energy, work, or quantity of heat, equivalent to meters squared kilograms per second squared. One joule is the work done when the point of application of a force of 1 newton is displaced a distance of 1 meter in the direction of the force (1 newton-meter).

kelvin (K) An SI unit of thermodynamic temperature. No degree sign (°) is used. Zero kelvin is absolute zero, the complete absence of heat.

kilo A prefix meaning 1,000.

kilobyte (kB) A unit of measurement for digital storage of data in various computer media, such as hard disks, random access memory, and compact discs. One kilobyte is 1,024 bytes.

kilograin A unit of weight equivalent to 1,000 grains.

kilogram (kg) An SI unit of mass. One kilogram is equal to the mass of a certain block of platinum–iridium alloy known as the international prototype kilogram (nicknamed Le Grand K), preserved at Sèvres, France. A new standard is expected early in the 21st century.

kilohertz (kHz) A unit of frequency equal to 1,000 hertz or 1,000 cycles per second.

kiloliter A unit of volume equal to 1,000 liters or 1 cubic meter.

kilopascal (kPa) A unit of pressure equal to 1,000 pascals.

kiloreactive volt-ampere (kvar) A unit of reactive power equal to 1,000 volt-ampere-reactive.

kilovolt (kV) A unit of electrical potential equal to 1,000 volts.

kilovolt-ampere (kVA) A unit of electrical power equal to 1,000 volt-amperes.

kilowatt (kW) A unit of electrical power equal to 1,000 watts.

kilowatt-hour (kW·hr) A unit of energy or work.

lambda (λ) A symbol used to represent 1 microliter. Avoid using this symbol; the preferred symbol is μL.

linear feet (lin ft) A unit of distance in feet along an object.

liter (L) A unit of volume. One liter of pure water weighs 1,000 grams at 4° Celsius at 1 atmosphere of pressure.

liters per day (L/day) A unit for expressing a volumetric flow rate past a given point.

liters per minute (L/min) A unit for expressing a volumetric flow rate past a given point.

lumen (lm) An SI unit of luminous flux equivalent to candela-steradian. One lumen is the luminous flux emitted in a solid angle of 1 steradian by a point source having a uniform intensity of 1 candela.

lux (lx) An SI unit of illuminance. One lux is the illuminance intensity given by a luminous flux of 1 lumen uniformly distributed over a surface of 1 square meter. One lux is equivalent to 1 candela-steradian per meter squared.

mega Prefix meaning 10^6 in Système International.

megabyte (MB) A unit of computer memory storage equivalent to 1,048,576 bytes.

megahertz (mHz) A unit of frequency equal to 1 million hertz, or 1 million cycles per second.

megaliter (ML) A unit of volume equal to 1 million liters.

megaohm (megohm) A unit of electrical resistance equal to 1 million ohms. This is the unit of measurement for testing the electrical resistance of water to determine its purity. The closer water comes to absolute purity, the greater its resistance to conducting an electric current. Absolutely pure water has a specific resistance of more than 18 million ohms across 1 centimeter at a temperature of 25° Celsius. See also *ohm*.

meter (m) An SI unit of length. One meter is the length of the path traveled by light in a vacuum during a time interval of 1/299,792,458 second.

meters per second per meter (m/sec/m; sec⁻¹) A unit for expressing velocity gradient.

metric system A system of units based on three basic units: the meter for length, the kilogram for mass, and the second for time—the so-called MKS system. Decimal fractions and multiples of the basic units are used for larger and smaller quantities. The principal departure of the SI from the more familiar form of metric engineering units is the use of the newton as the unit of force instead of kilogram-force. Likewise, the newton instead of kilogram-force is used in combination units including force; for example, pressure or stress (newton per square meter), energy (newton-meter = joule), and power (newton-meter per second = watt). See also *Système International*.

metric ton (t) A unit of weight equal to 1,000 kilograms.

mho A unit of electrical conductivity in US customary units equal to 1 siemens, which is an SI unit. See also *siemens*.

microgram (μg) A unit of mass equal to one millionth of a gram.

micrograms per liter (μg/L) A unit of concentration for dissolved substances based on their weights.

microhm A unit of electrical resistance equal to one millionth of an ohm.

micrometer (μm) A unit of length equal to one millionth of a meter.

micromho A unit of electrical conductivity equal to one millionth of an mho. See also *microsiemens*.

micromhos per centimeter (μmho/cm) A measure of the conductivity of a water sample, equivalent to microsiemens per centimeter. Absolutely pure water, from a mineral content standpoint, has a conductivity of 0.055 micromhos per centimeter at 25° Celsius.

micromolar (μ*M*) A concentration in which the molecular weight of a substance (in grams) divided by 10^6 (i.e., 1 μmol) is dissolved in enough solvent to make 1 liter of solution. See also *micromole; molar*.

micromole (μmol) A unit of weight for a chemical substance, equal to one millionth of a mole. See also *mole*.

micron (μ) A unit of length equal to 1 micrometer. Micrometers are the preferred units.

microsiemens (μS) A unit of conductivity equal to one millionth of a siemens. The microsiemens is the practical unit of measurement for conductivity and is used to approximate the total dissolved solids content of water. Water with 100 milligrams per liter of sodium chloride (NaCl) will have a specific resistance of 4,716 ohm-centimeters and a conductance of 212 microsiemens per centimeter. Absolutely pure water, from a mineral content standpoint, has a conductivity of 0.055 microsiemens per centimeter at 25° Celsius.

microwatt (μW) A unit of power equal to one millionth of a watt.

microwatt-seconds per square centimeter (μW-sec/cm^2) A unit of measurement of irradiation intensity and retention or contact time in the operation of ultraviolet systems.

mil A unit of length equal to one thousandth of an inch.

mile (mi) A unit of length, equivalent to 5,280 feet.

miles per hour (mph) A unit of speed.

milliampere (mA) A unit of electrical current equal to one thousandth of an ampere.

milliequivalent (meq) A unit of weight equal to one thousandth the equivalent weight of a chemical.

milliequivalents per liter (meq/L) A unit of concentration for dissolved substances based on their equivalent weights.

milligram (mg) A unit of mass equal to one thousandth of a gram.

milligrams per liter (mg/L) The unit used in reporting the concentration of matter in water as determined by water analyses.

milliliter (mL) A unit of volume equal to one thousandth of a liter.

millimeter (mm) A unit of length equal to one thousandth of a meter.

millimicron (mμ) A unit of length equal to one thousandth of a micron. This unit is correctly called a nanometer.

millimolar (mM) A concentration in which the molecular weight of a substance (in grams) divided by 10^3 (i.e., 1 mmol) is dissolved in enough solvent to make 1 liter of solution. See also *millimole; molar.*

millimole (mmol) A unit of weight for a chemical substance, equal to one-thousandth of a mole. See also *mole.*

million electron volts (MeV) A unit of energy equal to 10^6 electron volts. This unit is commonly used in the fields of nuclear and high-energy physics. See also *electron volt.*

million gallons (mil gal, MG) A unit of volume equal to 10^6.

million gallons per day (mgd) A unit for expressing the flow rate past a given point.

mils per year (mpy) A unit for expressing the loss of metal resulting from corrosion. Assuming the corrosion process is uniformly distributed over the test surface, the corrosion rate of a metal coupon may be converted to a penetration rate (length per time) by dividing the unit area of metal loss by the metal density (mass per volume). The penetration rate, expressed as mils per year, describes the rate at which the metal surface is receding because of the corrosion-induced metal loss. See also *mil.*

minute (min) A unit of time equal to 60 seconds.

molar (M) A unit for expressing the molarity of a solution. A 1-molar solution consists of 1 gram molecular weight of a compound dissolved in enough water to make 1 liter of solution. A gram molecular weight is the molecular weight of a compound in grams. For example, the molecular weight of sulfuric acid (H_2SO_4) is 98. A 1-molar, or 1-mole-per-liter, solution of sulfuric acid would consist of 98 grams of H_2SO_4 dissolved in enough distilled water to make 1 liter of solution.

mole (mol) A mole of a substance is a number of grams of that substance where the number equals the substance's molecular weight.

moles per liter (mol/L, M) A unit of concentration for a dissolved substance.

mrem An expression or measure of the extent of biological injury that would result from the absorption of a particular radionuclide at a given dosage over 1 year.

nanograms per liter (ng/L) A unit expressing the concentration of chemical constituents in solution as mass (nanograms) of solute per unit volume (liter) of water. One million nanograms per liter is equivalent to 1 milligram per liter.

nanometer (nm) A unit of length defined as 10^{-12} meter.

nephelometric turbidity unit (ntu) A unit for expressing the cloudiness (turbidity) of a sample as measured by a nephelometric turbidimeter. A turbidity of 1 nephelometric turbidity unit is equivalent to the turbidity created by a 1:4,000 dilution of a stock solution of 5.0 milliliters of a 1.000-gram hydrazine sulfate (($NH_2)_2 \cdot H_2SO_4$) in 100 milliliters of distilled water solution plus 5.0 milliliters of a 10.00-gram hexamethylene-tetramine (($CH_2)_6N_4$) in 100 milliliters of distilled water solution that has stood for 24 hours at $25 \pm 3°$ Celsius.

newton (N) An SI unit of force. One newton is equivalent to 1 kilogram-meter per second squared. It is the force, when applied to a body having a mass of 1 kilogram, that gives the body an acceleration of 1 meter per second squared. The newton replaces the unit kilogram-force, which is the unit of force in the metric system.

ohm (Ω) An SI unit of electrical resistance, equivalent to meters squared kilograms per second cubed per ampere squared. One ohm is the electrical resistance between two points of a conductor when a constant difference of 1 volt potential applied between the two points produces in the conductor a current of 1 ampere, with the conductor not being the source of any electromotive force.

one hundred cubic feet (ccf) A unit of volume.

ounce (oz) A unit of force, mass, and volume.

ounce-inch (ounce-in., ozf-in.) A unit of torque.

parts per billion (ppb) A unit of proportion, equal to 10^{-9}. This expression represents a measure of the concentration of a substance dissolved in water on a weight-per-weight basis or the concentration of a substance in air on a weight-per-volume basis. One liter of water at 4° Celsius has a mass equal to 1.000 kilogram (specific gravity equal to 1.000, or 1 billion micrograms). Thus, when 1 microgram of a substance is dissolved in 1 liter of water with a specific gravity of 1.000 (1 microgram per liter), this would be one part of substance per billion parts of water on a weight-per-weight basis. This terminology is now obsolete, and the term *micrograms per liter (µg/L)* should be used for concentrations in water.

parts per million (ppm) A unit of proportion, equal to 10^{-6}. This terminology is now obsolete, and the term *milligrams per liter (mg/L)* should be used for concentrations in water. See also *parts per billion*.

parts per thousand (ppt) A unit of proportion, equal to 10^{-3}. This terminology is now obsolete, and the term *grams per liter (g/L)* should be used for concentrations in water. See also *parts per billion*.

parts per trillion (ppt) A unit of proportion, equal to 10^{-12}. This terminology is now obsolete, and the term *nanograms per liter (ng/L)* should be used for concentrations in water. See also *parts per billion*.

pascal (Pa) An SI unit of pressure or stress equivalent to newtons per meter per second squared. One pascal is the pressure or stress of 1 newton per square meter.

pascal-second (Pa·sec) A unit of absolute viscosity equivalent to kilogram per second per meter cubed. The viscosity of pure water at 20° Celsius is 0.0010087 pascal-second.

pi (π) The ratio of the circumference of a circle to the diameter of that circle, approximately equal to 3.14159 (or about $^{22}/_{7}$).

picocurie (pCi) A unit of radioactivity. One picocurie represents a quantity of radioactive material with an activity equal to one millionth of one millionth of a curie (i.e., 10^{-12} curie).

picocuries per liter (pCi/L) A radioactivity concentration unit.

picogram (pg) A unit of mass equal to 10^{-12} gram or 10^{-15} kilogram.

picosecond (ps) A unit of time equal to one trillionth (10^{-12}) of a second.

plaque-forming unit (pfu) A unit expressing the number of infectious virus particles. One plaque-forming unit is equivalent to one virus particle.

platinum–cobalt (Pt–Co) color unit (PCU) See *color unit*.

poise A unit of absolute viscosity, equivalent to 1 gram mass per centimeter per second.

pound (lb) A unit used to represent either a mass or a force. This can be a confusing unit because two terms actually exist, *pound mass* (lbm) and *pound force* (lbf). One pound force is the force with which a 1-pound mass is attracted to the earth. In equation form,

$$\text{pounds force} = (\text{pounds mass})\left(\frac{\text{local acceleration resulting from gravity}}{\text{standard acceleration resulting from gravity}}\right)$$

One pound mass, on the other hand, is the mass that will accelerate at 32.2 feet per second squared when a 1-pound force is applied to it. As an example of the effect of the local acceleration resulting from gravity, at 10,000 feet (3,300 meters) above sea level, where the acceleration resulting from gravity is 32.17 feet per second squared (979.6 centimeters per second squared) instead of the sea level value of 32.2 feet per second squared (980.6 centimeters per second squared), the force of gravity on a 1-pound mass would be 0.999 pounds force. On the surface of the earth at sea level, pound mass and pound force are numerically the same because the acceleration resulting from gravity is applied to an object, although they are quite different physical quantities. This may lead to confusion.

pound force (lbf) See *pound.*

pound mass (lbm) See *pound.*

pounds per day (lb/day) A unit for expressing the rate at which a chemical is added to a water treatment process.

pounds per square foot (lb/ft^2) A unit of pressure.

pounds per square inch (psi) A unit of pressure.

pounds per square inch absolute (psia) A unit of pressure reflecting the sum of gauge pressure and atmospheric pressure.

pounds per square inch gauge (psig) A unit of pressure reflecting the pressure measured with respect to that of the atmosphere. The gauge is adjusted to read zero at the surrounding atmospheric pressure.

rad (radiation absorbed dose) A unit of adsorbed dose of ionizing radiation. Exposure of soft tissue or similar material to 1 roentgen results in the absorption of about 100 ergs (10^{-5} joules) of energy per gram, which is 1 rad. See also *gray; rem; sievert.*

radian (rad) An SI unit of measure of a plane angle that is equal to the angle at the center of a circle subtended by an arc equal in length to the radius. This unit is also used to measure the phase angle in a periodic electrical wave. Note that 2π radians is equivalent to $360°$.

radians per second (rad/sec) A unit of angular frequency.

rem (roentgen equivalent man [person]) A unit of equivalent dose of ionizing radiation, developed by the International Commission on Radiation Units and Measurements in 1962 to reflect the finding that the biological effects of ionizing radiation were dependent on the nature of the radiation as well as other factors. For X- and gamma radiation, the weighting factor is 1; thus, 1 rad equals 1 rem. For alpha radiation, however, 1 rad equals 20 rem. See also *gray*; *rad*; *sievert*.

revolutions per minute (rpm) A unit for expressing the frequency of rotation, or the number of times a fixed point revolves around its axis in 1 minute.

revolutions per second (rps) A unit for expressing the frequency of rotation, or the number of times a fixed point revolves around its axis in 1 second.

roentgen (r) The quantity of electrical charge produced by X- or gamma radiation. One roentgen of exposure will produce about 2 billion ion pairs per cubic centimeter of air. First introduced at the Radiological Congress held in Stockholm as the special unit for expressing exposure to ionizing radiation, it is now obsolete. See also *gray*; *rad*; *rem*; *sievert*.

second (sec) An SI unit of the duration of 9,192,631,770 periods of radiation corresponding to the transition between the two hyperfine levels of the ground state of the cesium-133 atom.

second feet A unit of flow equivalent to cubic feet per second.

second-foot day A unit of volume. One second-foot day is the discharge during a 24-hour period when the rate of flow is 1 second foot (i.e., 1 cubic foot per second). In ordinary hydraulic computations, 1 cubic foot per second flowing for 1 day is commonly taken as 2 acre-feet. The US Geological Survey now uses the term *cfs day* (cubic feet per second day) in its published reports.

section A unit of area in public land surveying. One section is a land area of 1 square mile.

SI See *Système International*.

siemens (S) An SI unit of the derived unit for electrical conductance, equivalent to seconds cubed amperes squared per meter squared per kilogram. One siemens is the electrical conductance of a conductor in which a current of 1 ampere is produced by an electric potential difference of 1 volt.

sievert (Sv) An SI unit of *equivalent* ionizing radiation dose. One sievert is the dose equivalent when the adsorbed dose of ionizing radiation multiplied by the dimensionless factors Q (quality factors) and N (product of any other multiplying factors) is 1 joule per kilogram. One sievert is equal to 100 rem. See also *gray*; *rad*; *rem*.

slug The base unit of mass. A slug is a mass that will accelerate at 1 foot per second squared when 1 pound force is applied.

square foot (ft^2) A unit of area equivalent to that of a square, 1 foot on each side.

square inch (in.^{2}) A unit of area equivalent to that of a square, 1 inch on each side.

square meter (m^2) A unit of area equivalent to that of a square, 1 meter on each side.

square mile (mi^2) A unit of area equivalent to that of a square, 1 mile on each side.

standard cubic feet per minute (SCFM) A unit for expressing the flow rate of air. This unit represents cubic feet of air per minute at standard conditions of temperature, pressure, and humidity (32° Fahrenheit, 14.7 pounds per square inch absolute, and 50% relative humidity).

steradian (sr) An SI unit of measure of a solid angle which, having its vertex in the center of a sphere, cuts off an area on the surface of the sphere equal to that of a square with sides of length equal to the radius of the sphere.

Système International (SI) The International System of Units of measure as defined by the periodic meeting of the General Conference on Weights and Measures. This system is sometimes called the international metric system or Le Système International d'Unités. The SI is a rationalized selection of units from the metric system with seven base units for which names, symbols, and precise definitions have been established. Many derived units are defined in terms of the base units, with symbols assigned to each and, in some cases, given names (e.g., the newton [N]). The great advantage of SI is its establishment of one and only one unit for each physical quantity—the meter for length, the kilogram (not the gram) for mass, the second for time, and so on. From these elemental units, units for all other mechanical quantities are derived. Another advantage is the ease with which unit conversions can be made, as few conversion factors need to be invoked.

tesla (T) An SI unit of magnetic flux density, equivalent to kilograms per second squared per ampere. One tesla is the magnetic flux density given by a magnetic flux of 1 weber per square meter.

ton A unit of force and mass defined as 2,000 pounds.

tonne (t) A unit of mass defined as 1,000 kilograms. A tonne is sometimes called a metric ton.

torr A unit of pressure. One torr is equal to 1 centimeter of mercury at 0° Celsius.

true color unit (tcu) A unit of color measurement based on the platinum–cobalt color unit. This unit is applied to water samples in which the turbidity has been removed. One true color unit equals 1 color unit. See also *color unit*.

turbidity unit See *nephelometric turbidity unit*.

US customary system of units A system of units based on the yard and the pound, commonly used in the United States and defined in "Unit of Weights and Measures (United States Customary and Metric): Definitions and Tables of Equivalents," *National Bureau of Standards Miscellaneous Publication MP 233*, Dec. 20, 1960. Most of the units have a historical origin from the United Kingdom (e.g., the length of a king's foot for the length of 1 foot; the area a team of horses could plow in a day—without getting tired—for an acre; the load a typical horse could lift in a minute for horsepower, and so forth). No organized method of multiples and fractions is involved. See also *Système International*.

volt (V) An SI unit of electrical potential, potential difference, and electromotive force, equivalent to meters squared kilograms per second cubed per ampere. One volt is the difference of electric potential between two points of a conductor, carrying a constant current of 1 ampere, when the power dissipated between these points is equal to 1 watt.

volt-ampere (VA) A unit used for expressing apparent power and complex power.

volt-ampere-reactive (VAR) A unit used for expressing reactive power.

watt (W) An SI unit of power and radiant flux, equivalent to meters squared kilograms per second cubed. One watt is the power that gives rise to the production of energy at the rate of 1 joule per second. Watts represent a measure of active power and instantaneous power.

weber (Wb) An SI unit of magnetic flux, equivalent to meters squared kilograms per second squared per ampere. One weber is the magnetic flux that, linking a circuit of one turn, produces in the circuit an electromotive force of 1 volt as the magnetic flux is reduced to zero at a uniform rate in 1 second.

yard (yd) A unit of length equal to 3 feet.

CONVERSION OF US CUSTOMARY UNITS _____

Linear Measurement

fathoms	× 6	= feet (ft)
feet (ft)	× 12	= inches (in.)
inches (in.)	× 0.0833	= feet (ft)
miles (mi)	× 5,280	= feet (ft)
yards (yd)	× 3	= feet (ft)
yards (yd)	× 36	= inches (in.)

Circular Measurement

degrees (angle)	× 60	= minutes (angle)
degrees (angle)	× 0.01745	= radians

Area Measurement

acres	× 43,560	= square feet (ft^2)
square feet (ft^2)	× 144	= square inches (in.^2)
square inches (in.^2)	× 0.00695	= square feet (ft^2)
square miles (mi^2)	× 640	= acres
square miles (mi^2)	× 27,880,000	= square feet (ft^2)
square miles (mi^2)	× 3,098,000	= square yards (yd^2)
square yards (yd^2)	× 9	= square feet (ft^2)

Volume Measurement

acre-feet (acre-ft)	× 43,560	= cubic feet (ft^3)
acre-feet (acre-ft)	× 325,851	= gallons (gal)
barrels (bbl)	× 42	= gallons (gal)
board foot (fbm)		= 144 square inches × 1 inch
cubic feet (ft^3)	× 1,728	= cubic inches (in.^3)
cubic feet (ft^3)	× 7.48052	= gallons (gal)
cubic feet (ft^3)	× 29.92	= quarts (qt)
cubic feet (ft^3)	× 59.84	= pints (pt)
cubic feet (ft^3)	× 0.000023	= acre feet (acre-ft)
cubic inches (in.^3)	× 0.00433	= gallons (gal)
cubic inches (in.^3)	× 0.00058	= cubic feet (ft^3)
drops	× 60	= teaspoons (tsp)
gallons (gal)	× 0.1337	= cubic feet (ft^3)
gallons (gal)	× 231	= cubic inches (in.^3)
gallons (gal)	× 0.0238	= barrels (bbl)
gallons (gal)	× 4	= quarts (qt)
gallons (gal)	× 8	= pints (pt)

gallons, US	× 0.83267	= gallons, Imperial
gallons (gal)	× 0.00000308	= acre-feet (acre-ft)
gallons (gal)	× 0.0238	= barrels (42 gal) (bbl)
gallons, Imperial	× 1.20095	= gallons, US
pints (pt)	× 2	= quarts (qt)
quarts (qt)	× 4	= gallons (gal)
quarts (qt)	× 57.75	= cubic inches (in.^{3})

Pressure Measurement

atmospheres	× 29.92	= inches of mercury
atmospheres	× 33.90	= feet of water
atmospheres	× 14.70	= pounds per square inch (lb/in.^{2})
feet of water	× 0.8826	= inches of mercury
feet of water	× 0.02950	= atmospheres
feet of water	× 0.4335	= pounds per square inch (lb/in.^{2})
feet of water	× 62.43	= pounds per square foot (lb/ft^2)
feet of water	× 0.8876	= inches of mercury
inches of mercury	× 1.133	= feet of water
inches of mercury	× 0.03342	= atmospheres
inches of mercury	× 0.4912	= pounds per square inch (lb/in.^{2})
inches of water	× 0.002458	= atmospheres
inches of water	× 0.07355	= inches of mercury
inches of water	× 0.03613	= pounds per square inch (lb/in.^{2})
pounds/square in. (lb/in.^{2})	× 0.01602	= feet of water
pounds/square foot (lb/ft^2)	× 6,954	= pounds per square inch (lb/in.^{2})
pounds/square in. (lb/in.^{2})	× 2.307	= feet of water
pounds/square inch (lb/in.^{2})	× 2.036	= inches of mercury
pounds/square inch (lb/in.^{2})	× 27.70	= inches of water
feet suction lift of water	× 0.882	= inches of mercury

Weight Measurement

cubic feet of ice	× 57.2	= pounds (lb)
cubic feet of water (50°F)	× 62.4	= pounds of water
cubic inches of water	× 0.036	= pounds of water
gallons of water (50°F)	× 8.3453	= pounds of water
milligrams/liter (mg/L)	× 0.0584	= grains per gallon (US) (gpg)
milligrams/liter (mg/L)	× 0.07016	= grains per gallon (Imperial)
milligrams/liter (mg/L)	× 8.345	= pounds per million gallons (lb/mil gal)
ounces (oz)	× 437.5	= grains (gr)

parts per million (ppm)	× 1	= milligrams per liter (mg/L) (for normal water applications)
grains per gallon (gpg)	× 17.118	= parts per million (ppm)
grains per gallon (gpg)	× 142.86	= pounds per million gallons (lb/mil gal)
percent solution	× 10,000	= milligrams per liter (mg/L)
pounds (lb)	× 16	= ounces (oz)
pounds (lb)	× 7,000	= grains (gr)
pounds (lb)	× 0.0004114	= tons (short)
pounds/cubic inch (lb/in.^{3})	× 1,728	= pounds per cubic foot (lb/ft^3)
pounds of water	× 0.0166032	= cubic feet (ft^3)
pounds of water	× 2,768	= cubic inches (in.^{3})
pounds of water	× 0.1198	= gallons (gal)
tons (short)	× 2,000	= pounds (lb)
tons (short)	× 0.89287	= tons (long)
tons (long)	× 2,240	= pounds (lb)
cubic feet air (at 60°F and 29.92 in. mercury)	× 0.0763	= pounds (lb)

Flow Measurement

barrels per hour (bbl/hr)	× 0.70	= gallons per minute (gpm)
acre-feet/minute	× 325.851	= gallons per minute (gpm)
acre-feet/minute	× 726	= cubic feet per second (ft^3/sec)
cubic feet/minute (ft^3/min)	× 0.1247	= gallons per second (gps)
cubic feet/minute (ft^3/min)	× 62.43	= pounds of water per minute
cubic feet/second (ft^3/sec)	× 448.831	= gallons per minute (gpm)
cubic feet/second (ft^3/sec)	× 0.646317	= million gallons per day (mgd)
cubic feet/second (ft^3/sec)	× 1.984	= acre-feet per day (acre-ft/day)
gallons/minute (gpm)	× 1,440	= gallons per day (gpd)
gallons/minute (gpm)	× 0.00144	= million gallons per day (mgd)
gallons/minute (gpm)	× 0.00223	= cubic feet per second (ft^3/sec)
gallons/minute (gpm)	× 0.1337	= cubic feet per minute (ft^3/min)
gallons/minute (gpm)	× 8.0208	= cubic feet per hour (ft^3/hr)
gallons/minute (gpm	× 0.00442	= acre-feet per day (acre-ft/day)
gallons/minute (gpm)	× 1.43	= barrels (42 gal) per day (bbl/day)
gallons water/minute	× 6.0086	= tons of water per 24 hours
million gallons/day (mgd)	× 1.54723	= cubic feet per second (ft^3/sec)
million gallons/day (mgd)	× 92.82	= cubic feet per minute (ft^3/min)
million gallons/day (mgd)	× 694.4	= gallons per minute (gpm)
million gallons/day (mgd)	× 3.07	= acre-feet per day (acre-ft/day)

pounds of water/minute	× 26.700	= cubic feet per second (ft³/sec)
miner's inch		= flow through an orifice of 1 in.² under a head of 4 to 6 in.
miner's inches (9 gpm)	× 8.98	= gallons per minute (gpm)
miner's inches (9 gpm)	× 1.2	= cubic feet per minute (ft³/min)
miner's inches (11.25 gpm)	× 11.22	= gallons per minute (gpm)
miner's inches (11.25 gpm)	× 1.5	= cubic feet per minute (ft³/min)

Work Measurement

British thermal units (Btu)	× 777.5	= foot-pounds (ft-lb)
British thermal units (Btu)	× 39,270	= horsepower-hours (hp·hr)
British thermal units (Btu)	× 29,280	= kilowatt-hours (kW·hr)
foot-pounds (ft-lb)	× 1,286	= British thermal units (Btu)
foot-pounds (ft-lb)	× 50,500,000	= horsepower-hours (hp·hr)
foot-pounds (ft-lb)	× 37,660,000	= kilowatt-hours (kW·hr)
horsepower-hours (hp·hr)	× 2,547	= British thermal units (Btu)
horsepower-hours (hp·hr)	× 0.7457	= kilowatt-hours (kW·hr)
kilowatt-hours (kW·hr)	× 3,415	= British thermal units (Btu)
kilowatt-hours (kW·hr)	× 1.241	= horsepower-hours (hp·hr)

Power Measurement

boiler horsepower	× 33,480	= British thermal units per hour (Btu/hr)
boiler horsepower	× 9.8	= kilowatts (kW)
British thermal units/second (Btu/sec)	× 1.0551	= kilowatts (kW)
British thermal units/minute (Btu/min)	× 12.96	= foot-pounds per second (ft-lb/sec)
British thermal units/minute (Btu/min)	× 0.02356	= horsepower (hp)
British thermal units/minute (Btu/min)	× 0.01757	= kilowatts (kW)
British thermal units/hour (Btu/hr)	× 0.293	= watts (W)
British thermal units/hour (Btu/hr)	× 12.96	= foot-pounds per minute (ft-lb/min)
British thermal units/hour (Btu/hr)	× 0.00039	= horsepower (hp)
foot-pounds per second (ft-lb/sec)	× 771.7	= British thermal units per minute (Btu/min)

Units of Measure and Conversions

foot-pounds per second (ft-lb/sec)	× 1,818	= horsepower (hp)
foot-pounds per second (ft-lb/sec)	× 1,356	= kilowatts (kW)
foot-pounds per minute (ft-lb/min)	× 303,000	= horsepower (hp)
foot-pounds per minute (ft-lb/min)	× 226,000	= kilowatts (kW)
horsepower (hp)	× 42.44	= British thermal units per minute (Btu/min)
horsepower (hp)	× 33,000	= foot-pounds per minute (ft-lb/min)
horsepower (hp)	× 550	= foot-pounds per second (ft-lb/sec)
horsepower (hp)	× 1,980,000	= foot-pounds per hour (ft-lb/hr)
horsepower (hp)	× 0.7457	= kilowatts (kW)
horsepower (hp)	× 745.7	= watts (W)
kilowatts (kW)	× 0.9478	= British thermal units per second (Btu/sec)
kilowatts (kW)	× 56.92	= British thermal units per minute (Btu/min)
kilowatts (kW)	× 3,413	= British thermal units per hour (Btu/hr)
kilowatts (kW)	× 44,250	= foot-pounds per minute (ft-lb/min)
kilowatts (kW)	× 737.6	= foot-pounds per second (ft-lb/sec)
kilowatts (kW)	× 1.341	= horsepower (hp)
tons of refrigeration (US)	× 288,000	= British thermal units per 24 hours
watts (W)	× 0.05692	= British thermal units per minute (Btu/min)
watts (W)	× 0.7376	= foot-pounds (force) per second (ft-lb/sec)
watts (W)	× 44.26	= foot-pounds per minute (ft-lb/min)
watts (W)	× 1,341	= horsepower (hp)

Velocity Measurement

feet/minute (ft/min)	× 0.01667	= feet per second (ft/sec)
feet/minute (ft/min)	× 0.01136	= miles per hour (mph)
feet/second (ft/sec)	× 0.6818	= miles per hour (mph)

| miles/hour (mph) | × 88 | = feet per minute (ft/min) |
| miles/hour (mph) | × 1.467 | = feet per second (ft/sec) |

Miscellaneous

| grade: 1 percent (or 0.01) | | = 1 foot per 100 feet |

CONVERSION OF METRIC UNITS

Linear Measurement

inch (in.)	× 25.4	= millimeters (mm)
inch (in.)	× 2.54	= centimeters (cm)
foot (ft)	× 304.8	= millimeters (mm)
foot (ft)	× 30.48	= centimeters (cm)
foot (ft)	× 0.3048	= meters (m)
yard (yd)	× 0.9144	= meters (m)
mile (mi)	× 1,609.3	= meters (m)
mile (mi)	× 1.6093	= kilometers (km)
millimeter (mm)	× 0.03937	= inches (in.)
centimeter (cm)	× 0.3937	= inches (in.)
meter (m)	× 39.3701	= inches (in.)
meter (m)	× 3.2808	= feet (ft)
meter (m)	× 1.0936	= yards (yd)
kilometer (km)	× 0.6214	= miles (mi)

Area Measurement

square meter (m^2)	× 10,000	= square centimeters (cm^2)
hectare (ha)	× 10,000	= square meters (m^2)
square inch (in.^{2})	× 6.4516	= square centimeters (cm^2)
square foot (ft^2)	× 0.092903	= square meters (m^2)
square yard (yd^2)	× 0.8361	= square meters (m^2)
acre	× 0.004047	= square kilometers (km^2)
acre	× 0.4047	= hectares (ha)
square mile (mi^2)	× 2.59	= square kilometers (km^2)
square centimeter (cm^2)	× 0.16	= square inches (in.^{2})
square meters (m^2)	× 10.7639	= square feet (ft^2)
square meters (m^2)	× 1.1960	= square yards (yd^2)
hectare (ha)	× 2.471	= acres
square kilometer (km^2)	× 247.1054	= acres
square kilometer (km^2)	× 0.3861	= square miles (mi^2)

Units of Measure and Conversions

Volume Measurement

cubic inch (in.^{3})	× 16.3871	= cubic centimeters (cm^3)
cubic foot (ft^3)	× 28,317	= cubic centimeters (cm^3)
cubic foot (ft^3)	× 0.028317	= cubic meters (m^3)
cubic foot (ft^3)	× 28.317	= liters (L)
cubic yard (yd^3)	× 0.7646	= cubic meters (m^3)
acre foot (acre-ft)	× 123.34	= cubic meters (m^3)
ounce (US fluid) (oz)	× 0.029573	= liters (L)
quart (liquid) (qt)	× 946.9	= milliliters (mL)
quart (liquid) (qt)	× 0.9463	= liters (L)
gallon (gal)	× 3.7854	= liters (L)
gallon (gal)	× 0.0037854	= cubic meters (m^3)
peck (pk)	× 0.881	= decaliters (dL)
bushel (bu)	× 0.3524	= hectoliters (hL)
cubic centimeters (cm^3)	× 0.061	= cubic inches (in.^{3})
cubic meter (m^3)	× 35.3183	= cubic feet (ft^3)
cubic meter (m^3)	× 1.3079	= cubic yards (yd^3)
cubic meter (m^3)	× 264.2	= gallons (gal)
cubic meter (m^3)	× 0.000811	= acre-feet (acre-ft)
liter (L)	× 1.0567	= quart (liquid) (qt)
liter (L)	× 0.264	= gallons (gal)
liter (L)	× 0.0353	= cubic feet (ft^3)
decaliter (dL)	× 2.6417	= gallons (gal)
decaliter (dL)	× 1.135	= pecks (pk)
hectoliter (hL)	× 3.531	= cubic feet (ft^3)
hectoliter (hL)	× 2.84	= bushels (bu)
hectoliter (hL)	× 0.131	= cubic yards (yd^3)
hectoliter (hL)	× 26.42	= gallons (gal)

Pressure Measurement

pound/square inch (psi)	× 6.8948	= kilopascals (kPa)
pound/square inch (psi)	× 0.00689	= pascals (Pa)
pound/square inch (psi)	× 0.070307	= kilograms/square centimeter (kg/cm^2)
pound/square foot (lb/ft^2)	× 47.8803	= pascals (Pa)
pound/square foot (lb/ft^2)	× 0.000488	= kilograms/square centimeter (kg/cm^2)

pound/square foot (lb/ft^2)	× 4.8824	= kilograms/square meter (kg/m^2)
inches of mercury	× 3,376.8	= pascals (Pa)
inches of water	× 248.84	= pascals (Pa)
bar	× 100,000	= newtons per square meter (N/m^2)
pascals (Pa)	× 1	= newtons per square meter (N/m^2)
pascals (Pa)	× 0.000145	= pounds/square inch (psi)
kilopascals (kPa)	× 0.145	= pounds/square inch (psi)
pascals (Pa)	× 0.000296	= inches of mercury (at 60°F)
kilogram/square centimeter (kg/cm^2)	× 14.22	= pounds/square inch (psi)
kilogram/square centimeter (kg/cm^2)	× 28.959	= inches of mercury (at 60°F)
kilogram/square meter (kg/m^2)	× 0.2048	= pounds per square foot (lb/ft^2)
centimeters of mercury	× 0.4461	= feet of water

Weight Measurement

ounce (oz)	× 28.3495	= grams (g)
pound (lb)	× 0.045359	= grams (g)
pound (lb)	× 0.4536	= kilograms (kg)
ton (short)	× 0.9072	= megagrams (metric ton)
pounds/cubic foot (lb/ft^3)	× 16.02	= grams per liter (g/L)
pounds/million gallons (lb/mil gal)	× 0.1198	= grams per cubic meter (g/m^3)
gram (g)	× 15.4324	= grains (gr)
gram (g)	× 0.0353	= ounces (oz)
gram (g)	× 0.0022	= pounds (lb)
kilograms (kg)	× 2.2046	= pounds (lb)
kilograms (kg)	× 0.0011	= tons (short)
megagram (metric ton)	× 1.1023	= tons (short)
grams/liter (g/L)	× 0.0624	= pounds per cubic foot (lb/ft^3)
grams/cubic meter (g/m^3)	× 8.3454	= pounds/million gallons (lb/mil gal)

Flow Measurement

gallons/second (gps)	× 3.785	= liters per second (L/sec)
gallons/minute (gpm)	× 0.00006308	= cubic meters per second (m^3/sec)
gallons/minute (gpm)	× 0.06308	= liters per second (L/sec)

gallons/hour (gph)	× 0.003785	= cubic meters per hour (m^3/hr)
gallons/day (gpd)	× 0.000003785	= million liters per day (ML/day)
gallons/day (gpd)	× 0.003785	= cubic meters per day (m^3/day)
cubic feet/second (ft^3/sec)	× 0.028317	= cubic meters per second (m^3/sec)
cubic feet/second (ft^3/sec)	× 1,699	= liters per minute (L/min)
cubic feet/minute (ft^3/min)	× 472	= cubic centimeters/second (cm^3/sec)
cubic feet/minute (ft^3/min)	× 0.472	= liters per second (L/sec)
cubic feet/minute (ft^3/min)	× 1.6990	= cubic meters per hour (m^3/hr)
million gallons/day (mgd)	× 43.8126	= liters per second (L/sec)
million gallons/day (mgd)	× 0.003785	= cubic meters per day (m^3/day)
million gallons/day (mgd)	× 0.043813	= cubic meters per second (m^3/sec)
gallons/square foot (gal/ft^2)	× 40.74	= liters per square meter (L/m^2)
gallons/acre/day (gal/acre/day)	× 0.0094	= cubic meters/hectare/day (m^3/ha/day)
gallons/square foot/day (gal/ft^2/day)	× 0.0407	= cubic meters/square meter/day (m^3/m^2/day)
gallons/square foot/day (gal/ft^2/day)	× 0.0283	= liters/square meter/day (L/m^2/day)
gallons/square foot/minute (gal/ft^2/min)	× 2.444	= cubic meters/square meter/hour (m^3/m^2/hr) = m/hr
gallons/square foot/minute (gal/ft^2/min)	× 0.679	= liters/square meter/second (L/m^2/sec)
gallons/square foot/minute (gal/ft^2/min)	× 40.7458	= liters/square meter/minute (L/m^2/min)
gallons/capita/day (gpcd)	× 3.785	= liters/day/capita (L/d/capita)
liters/second (L/sec)	× 22,824.5	= gallons per day (gpd)
liters/second (L/sec)	× 0.0228	= million gallons per day (mgd)
liters/second (L/sec)	× 15.8508	= gallons per minute (gpm)
liters/second (L/sec)	× 2.119	= cubic feet per minute (ft^3/min)
liters/minute (L/min)	× 0.0005886	= cubic feet per second (ft^3/sec)
cubic centimeters/second (cm^3/sec)	× 0.0021	= cubic feet per minute (ft^3/min)
cubic meters/second (m^3/sec)	× 35.3147	= cubic feet per second (ft^3/sec)
cubic meters/second (m^3/sec)	× 22.8245	= million gallons per day (mgd)
cubic meters/second (m^3/sec)	× 15,850.3	= gallons per minute (gpm)

cubic meters/hour (m³/hr)	× 0.5886	= cubic feet per minute (ft³/min)
cubic meters/hour (m³/hr)	× 4.403	= gallons per minute (gpm)
cubic meters/day (m³/day)	× 264.1720	= gallons per day (gpd)
cubic meters/day (m³/day)	× 0.00026417	= million gallons per day (mgd)
cubic meters/hectare/day (m³/ha/day)	× 106.9064	= gallons per acre per day (gal/acre/day)
cubic meters/square meter/day (m³/m²/day)	× 24.5424	= gallons/square foot/day (gal/ft²/day)
liters/square meter/minute (L/m²/min)	× 0.0245	= gallons/square foot/minute (gal/ft²/min)
liters/square meter/minute (L/m²/min)	× 35.3420	= gallons/square foot/day (gal/ft²/day)

Work, Heat, and Energy Measurements

British thermal units (Btu)	× 1.0551	= kilojoules (kJ)
British thermal units (Btu)	× 0.2520	= kilogram-calories (kg-cal)
foot-pound (force) (ft-lb)	× 1.3558	= joules (J)
horsepower-hour (hp·hr)	× 2.6845	= megajoules (MJ)
watt-second (W-sec)	× 1.000	= joules (J)
watt-hour (W·hr)	× 3.600	= kilojoules (kJ)
kilowatt-hour (kW·hr)	× 3,600	= kilojoules (kJ)
kilowatt-hour (kW·hr)	× 3,600,000	= joules (J)
British thermal units per pound (Btu/lb)	× 0.5555	= kilogram-calories per kilogram (kg-cal/kg)
British thermal units per cubic foot (Btu/ft³)	× 8.8987	= kilogram-calories/cubic meter (kg-cal/m³)
kilojoule (kJ)	× 0.9478	= British thermal units (Btu)
kilojoule (kJ)	× 0.00027778	= kilowatt-hours (kW·hr)
kilojoule (kJ)	× 0.2778	= watt-hours (W·hr)
joule (J)	× 0.7376	= foot-pounds (ft-lb)
joule (J)	× 1.0000	= watt-seconds (W-sec)
joule (J)	× 0.2399	= calories (cal)
megajoule (MJ)	× 0.3725	= horsepower-hour (hp·hr)
kilogram-calories (kg-cal)	× 3.9685	= British thermal units (Btu)
kilogram-calories per kilogram (kg-cal/kg)	× 1.8000	= British thermal units per pound (Btu/lb)
kilogram-calories per liter (kg-cal/L)	× 112.37	= British thermal units per cubic foot (Btu/ft³)

kilogram-calories/cubic meter (kg-cal/m^3)	× 0.1124	= British thermal units per cubic foot (Btu/ft^3)

Velocity, Acceleration, and Force Measurements

feet per minute (ft/min)	× 18.2880	= meters per hour (m/hr)
feet per hour (ft/hr)	× 0.3048	= meters per hour (m/hr)
miles per hour (mph)	× 44.7	= centimeters per second (cm/sec)
miles per hour (mph)	× 26.82	= meters per minute (m/min)
miles per hour (mph)	× 1.609	= kilometers per hour (km/hr)
feet/second/second (ft/sec^2)	× 0.3048	= meters/second/second (m/sec^2)
inches/second/second (in./sec^2)	× 0.0254	= meters/second/second (m/sec^2)
pound-force (lbf)	× 4.44482	= newtons (N)
centimeters/second (cm/sec)	× 0.0224	= miles per hour (mph)
meters/second (m/sec)	× 3.2808	= feet per second (ft/sec)
meters/minute (m/min)	× 0.0373	= miles per hour (mph)
meters per hour (m/hr)	× 0.0547	= feet per minute (ft/min)
meters per hour (m/hr)	× 3.2808	= feet per hour (ft/hr)
kilometers/second (km/sec)	× 2.2369	= miles per hour (mph)
kilometers/hour (km/hr)	× 0.0103	= miles per hour (mph)
meters/second/second (m/sec^2)	× 3.2808	= feet/second/second (ft/sec^2)
meters/second/second (m/sec^2)	× 39.3701	= inches/second/second (in./sec^2)
newtons (N)	× 0.2248	= pounds force (lbf)

Factors for Conversion

US	Multiply by	Metric (SI) or US
Length		
inches (in.)	2.540	centimeters (cm)
	0.0254	meters (m)
feet (ft)	0.3048	meter (m)
	12	inches (in.)
yard (yd)	0.9144	meters (m)
	3	feet (ft)
miles (mi)	1.609	kilometers (km)
	1,760	yards (yd)
	5,280	feet (ft)
Area		
square inch (in.^{2})	6.452	square centimeters (cm^2)
square feet (ft^2)	0.0929	square meters (m^2)
	144	square inches (in.^{2})
acres	4,047	square meters (m^2)
	0.4047	hectares (ha)
	43,560	square feet (ft^2)
	0.001562	square miles (mi^2)
square miles (mi^2)	2.590	square kilometers (km^2)
	640	acres
Volume		
cubic feet (ft^3)	28.32	liters (L)
	0.02832	cubic meters (m^3)
	7.48	gallons, US
	6.23	gallons, Imperial
	1,728	cubic inches (in.^{3})
cubic yard (yd^3)	0.7646	cubic meters (m^3)
gallon (gal)	3.785	liters (L)
	0.003785	cubic meters (m^3)
	4	quarts (qt)
	8	pints (pt)
	128	fluid ounces (fl oz)
	0.1337	cubic feet (ft^3)

Table continued on next page

Units of Measure and Conversions

Factors for Conversion (continued)

US	Multiply by	Metric (SI) or US
quarts (qt)	32	fl oz
	946	milliliters (mL)
	0.946	liters (L)
acre-feet (acre-ft)	1.233×10^{-3}	cubic hectometers (hm^3)
	1,233	cubic meters (m^3)
	1,613.3	cubic yd (yd^3)
Weight		
pounds (lb)	453.6	grams (g)
	0.4536	kilograms (kg)
	7,000	grains (gr)
	16	ounces (oz)
grains (gr)	0.0648	grams (g)
tons (short)	2,000	pounds (lb)
	0.9072	tonnes (metric tons)
tons (long)	2,240	pounds (lb)
gallons of water, US	8.34	pounds (lb)
gallons, Imperial	10	pounds (lb)
Unit Weight		
cubic feet (ft^3) of water	62.4	pounds (lb)
	7.48	gallons (gal)
pounds per cubic foot (lb/ft^3)	157.09	newtons per cubic meter (N/m^3)
	16.02	kilograms force/square meter (kgf/m^2)
	0.016	grams/cubic centimeter (g/cm^3)
pounds per ton	0.5	kilograms/metric ton (kg/tonne)
	0.5	milligrams per kilogram (mg/kg)
Concentration		
parts per million (ppm)	1	milligrams per liter (mg/L)
	8.34	pounds/million gallons (lb/mil gal)
grains per gallon (gpg)	17.4	milligrams per liter (mg/L)
	142.9	pounds/million gallons (lb/mil gal)
Time		
days	24	hours (hr)
	1,440	minutes (min)
	86,400	seconds (sec)

Table continued on next page

Factors for Conversion (continued)

US	Multiply by	Metric (SI) or US
hours (hr)	60	minutes (min)
minutes (min)	60	seconds (sec)
Slope		
feet per mile (ft/mi)	0.1894	meters per kilometer (m/km)
Velocity		
feet per second (ft/sec)	720	inches per minute (in./min)
	0.3048	meters per second (m/sec)
	30.48	centimeters per second (cm/sec)
	0.6818	miles per hour (mph)
inches per minute (in./min)	0.043	centimeters per second (cm/sec)
miles per hour (mph)	0.4470	meters per second (m/sec)
	1.609	kilometers per hour (km/hr)
knots	0.5144	meters per second (m/sec)
	1.852	kilometers per hour (km/hr)
Discharge		
cubic feet per second (ft^3/sec)	0.646	million gallons per day (mgd)
	448.8	gallons per minute (gpm)
	28.32	liters per second (L/sec)
	0.02832	cubic meters per second (m^3/sec)
million gallons per day (mgd)	3,785	metric tons per day
	3,785	cubic meters per day (m^3/day)
	0.04381	cubic meters per second (m^3/sec)
	694	gallons per minute (gpm)
	1.547	cubic feet per second (ft^3/sec)
gallons per minute (gpm)	3.785	liters per minute (L/min)
	0.06308	liters per second (L/sec)
	0.0000631	cubic meters per second (m^3/sec)
	8.021	cubic feet per hour (ft^3/hr)
	0.002228	cubic feet per second (ft^3/sec)
gallons per day (gpd)	3.785	liters (or kilograms) per day (L/day)
million gallons per day per acre-foot (mgd/acre-ft)	0.430	gallons per minute per cubic yard (gpm/yd^3)
	0.9354	cubic meters/square meter/day (m^3/m^2·day)

Table continued on next page

Units of Measure and Conversions

Factors for Conversion (continued)

US	Multiply by	Metric (SI) or US
acre-feet per day	0.01428	cubic meters/second (m³/sec)
Application Rate		
cubic feet per gallon (ft³/gal)	7.4805	cubic meters/cubic meter (m³/m³)
cubic feet per million gallons (ft³/mil gal)	0.00748	cubic meters/1,000 cubic meters (m³/1,000 m³)
cubic feet per 1,000 cubic feet per minute (ft³/1,000 ft³·min)	0.001	cubic meters/cubic meter/minute (m³/m³·min)
cubic feet per cubic feet per hour (ft³/ft³·hr)	180	gallons/square foot/day (gal/ft²·day)
cubic feet per minute per foot (ft³/min·ft)	0.00748	cubic meters per minute per meter (m³/min·m)
cubic feet/pound (ft³/lb)	0.0625	cubic meters per kilogram (m³/kg)
gallons per foot per day (gal/ft·day)	0.0124	cubic meters per meter per day (m³/m·day)
gallons per square foot per minute (gal/ft²·min)	40.7458	liters per square meter per minute (L/m²·min)
	0.04075	cubic meters per square meters per minute (m³/m²·min)
	2.445	cubic meters per square meter per hour (m³/m²·hr)
	58.6740	cubic meters per square meter per day (m³/m²·day)
gallons per acre (gal/acre)	0.00935	cubic meters per hectare (m³/ha)
million gallons per day per acre-foot (mgd/acre-ft)	0.430	gallons per minute per cubic yard (gpm/yd³)
pounds per acre (lb/acre)	1.121	kilograms per hectare (kg/ha)
pounds per pound per day (lb/lb·day)	1	kilograms per kilogram per day (kg/kg·day)
pounds per day (lb/day)	0.4536	kilograms per day (kg/day)
pounds per square foot per hour (lb/ft²·hr)	4.8827	kilograms per square meter per hour (kg/m²·hr)

Table continued on next page

Factors for Conversion (continued)

US	Multiply by	Metric (SI) or US
pounds per 1,000 square feet per day (lb/1,000 ft^2·day)	0.0049	kilograms per square meter per day (kg/m^2·day)
pounds per acre per day (lb/acre·day)	1.1209	kilograms per hectare per day (kg/ha·day)
pounds per cubic feet per hour (lb/ft^3·hr)	16.0185	kilograms per cubic meter per hour (kg/m^3·hr)
pounds per 1,000 cubic feet per day (lb/1,000 ft^3·day)	0.0160	kilograms per cubic meter per day (kg/m^3·day)
pounds per 1,000 gallons (lb/1,000 gal)	120.48	kilograms per 1,000 cubic meters (kg/1,000 m^3)
pounds per million gallons (lb/mil gal)	0.12	milligrams per liter (mg/L)
Force		
pounds (lb)	0.4536	kilograms force (kgf)
	453.6	grams (g)
	4.448	newtons (N)
Pressure		
pounds per square inch (psi)	2.309	feet head of water
	2.036	inches head of mercury
	51.71	millimeters of mercury
	6894.76	newtons per square meter (N/m^2) = pascals (Pa)
	703.1	kilograms of force per square meter (kgf/m^2)
	0.0690	bars
pounds per square foot (lb/ft^2)	4.882	kilograms of force per square meter (kgf/m^2)
	47.88	newtons per square meter (N/m^2)
pounds per cubic inch (lb/in.^{3})	0.01602	grams of force per cubic centimeter (gmf/cm^3)
	16	grams of force per liter (gmf/L)
tons per square inch	1.5479	kilograms per square millimeter (kg/mm^2)
millibars (mb)	100	newtons per square meter (N/m^2)

Table continued on next page

Factors for Conversion (continued)

US	Multiply by	Metric (SI) or US
inches of mercury	345.34	kilograms per square meter (kg/m^2)
	0.0345	kilograms per square centimeter (kg/cm^2)
	0.0334	bars
	0.491	per square inch (psi)
atmospheres	101,325	pascals (Pa)
	1,013	millibars (1 mb = 100 Pa)
	14.696	per square inch (psi)
pascals	1.0	newtons per square meter (N/m^2)
	1.0×10^{-5}	bars
	1.0200×10^{-5}	kilograms per square meter (kg/m^2)
	9.8692×10^{-6}	atmospheres
	1.40504×10^{-4}	per square inch (psi)
	4.0148×10^{-3}	inch head of water
	7.5001×10^{-4}	centimeters head of mercury
Mass and Density		
slugs	14.594	kilograms (kg)
	32.174	pound (lb) (mass)
pounds	0.4536	kilograms (kg)
slugs per cubic foot	515.4	kilograms per cubic meter (kg/m^3)
density (γ) of water	62.4	pounds per cubic meter (lb/ft^3) at 50°F
	980.2	newtons per cubic meter (N/m^3) at 10°C
specific weight (ρ) of water	1.94	slugs per cubic foot (slugs/ft^3)
	1,000	kilograms per cubic meter (kg/m^3)
	1	kilograms per liter (kg/L)
	1	grams per milliliter (g/mL)
Viscosity		
pound-seconds per cubic foot or slugs per foot-second (lb-sec/ft^3 or slugs/ft-sec)	47.88	newton-seconds per square meter (N-sec/m^2)
square feet per second (ft^2/sec)	0.0929	square meter per second (m^2/sec)

Table continued on next page

Factors for Conversion (continued)

US	Multiply by	Metric (SI) or US
Work		
British thermal units (Btu)	778	foot-pounds (ft-lb)
	0.293	watt-hours (W·hr)
	1	heat required to change 1 lb of water by 1°F
British thermal units/ pound (Btu/lb)	2.3241	kilojoules per kilogram (kJ/kg)
British thermal units/ cubic foot/degrees Fahrenheit/hour ($Btu/ft^3 \cdot °F \cdot hr$)	5.6735	watts per square meter per degrees Celsius per hour ($W/m^2 \cdot °C \cdot hr$)
horsepower-hours (hp·hr)	2,545	British thermal units (Btu)
	0.746	kilowatt-hours (kW·hr)
kilowatt-hours (kW·hr)	3,413	British thermal units (Btu)
	1.34	horsepower-hour (hp·hr)
horsepower per 1,000 gallons (hp/1,000 gal)	0.1970	kilowatts per cubic meter (kW/m^3)
Power		
horsepower (hp)	550	foot-pounds per second (ft-lb/sec)
	746	watts (W)
	2,545	British thermal units per hour (Btu/hr)
kilowatts (kW)	3,413	British thermal units per hour (Btu/hr)
British thermal units/hour (Btu/hr)	0.293	watts (W)
	12.96	foot-pounds per minute (ft-lb/min)
	0.00039	horsepower (hp)
Temperature		
degrees Fahrenheit (°F)	(°F − 32) × (5/9)	degrees Celsius (°C)
degrees Celsius (°C)	°C × (9/5) + 32	degrees Fahrenheit (°F)
	°C + 273.15	Kelvin (K)

Units of Measure and Conversions

Decimal Equivalents of Fractions

Fraction	Decimal	Fraction	Decimal
1/64	0.01563	33/64	0.51563
1/32	0.03125	17/32	0.53125
3/64	0.04688	35/64	0.54688
1/16	0.06250	9/16	0.56250
5/64	0.07813	37/64	0.57813
3/32	0.09375	19/32	0.59375
7/64	0.10938	39/64	0.60938
1/8	0.12500	5/8	0.62500
9/64	0.14063	41/64	0.64063
5/32	0.15625	21/32	0.65625
11/64	0.17188	43/64	0.67188
3/16	0.18750	11/16	0.68750
13/64	0.20313	45/64	0.70313
7/32	0.21875	23/32	0.71875
15/64	0.23438	47/64	0.73438
1/4	0.25000	3/4	0.75000
17/64	0.26563	49/64	0.76563
9/32	0.28125	25/32	0.78125
19/64	0.29688	51/64	0.79688
10/32	0.31250	13/16	0.81250
21/64	0.32813	53/64	0.82813
11/32	0.34375	27/32	0.84375
23/64	0.35938	55/64	0.85938
3/8	0.37500	7/8	0.87500
25/64	0.39063	57/64	0.89063
13/32	0.40625	29/32	0.90625
27/64	0.42188	59/64	0.92188
7/16	0.43750	15/16	0.93750
29/64	0.45313	61/64	0.95313
15/32	0.46875	31/32	0.96875
31/64	0.48438	63/64	0.98438
1/2	0.50000		

TEMPERATURE CONVERSIONS

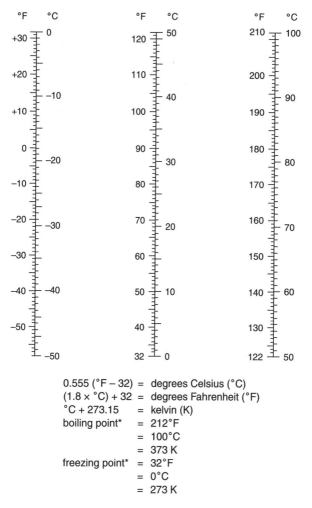

$$0.555 \, (°F - 32) = \text{degrees Celsius (°C)}$$
$$(1.8 \times °C) + 32 = \text{degrees Fahrenheit (°F)}$$
$$°C + 273.15 = \text{kelvin (K)}$$

$$\begin{aligned}
\text{boiling point*} &= 212°F \\
&= 100°C \\
&= 373 \text{ K} \\
\text{freezing point*} &= 32°F \\
&= 0°C \\
&= 273 \text{ K}
\end{aligned}$$

*At 14.696 psia, 101.325 kPa.

Celsius/Fahrenheit Comparison Graph

WATER CONVERSIONS

Water is composed of two gases, hydrogen and oxygen, in the ratio of two volumes of the former to one of the latter. It is never found pure in nature because of the readiness with which it absorbs impurities from the air and soil.

- One foot of water column at 39.1°F = 62.425 pounds on the square foot.

- One foot of water column at 39.1°F = 0.4335 pound on the square inch.

- One foot of water column at 39.1°F = 0.0295 atmospheric pressure.

- One foot of water column at 39.1°F = 0.8826 inch mercury column at 32°F.

- One foot of water column at 39.1°F = 773.3 feet of air column at 32°F and atmospheric pressure.

- One pound pressure per square foot = 0.01602 foot water column at 39.1°F.

- One pound pressure per square foot = 2.307 feet water column at 39.1°F.

- One atmospheric pressure = 29.92 inches mercury column = 33.9 feet water column.

- One inch of mercury column at 32°F = 1.133 feet water column.

- One foot of air column at 32°F and 1 atmospheric pressure = 0.001293 foot water column.

WATER EQUIVALENTS AND DATA

- 1 US gallon of water weighs 8.345 pounds.

- 1 cubic foot of water equals 7.48 gallons.

- 1 foot head of water develops 0.433 pounds pressure per square inch.

- Pounds per hour times 0.12 equals gallons per hour.

- Grains per gallon times 0.143 equals pounds per 1,000 gallons.

- Parts per million divided by 120 equals pounds per 1,000 gallons.

- 1 grain per gallon equals 17.1 parts per million.

- Estimated flow in gallons per minute equals pipe diameter in inches squared times 20.

- 1 boiler horsepower based on 10 square feet of heating surface requires 4 gallons per hour of feedwater.

- 1 pound of coal will produce 7 to 10 pounds of steam.

- 1 gallon of oil will produce 70 to 120 pounds of steam.

- 1,000 cubic feet of natural gas will produce 600 pounds of steam.

- Saturated salt brine for zeolite regeneration contains 2.48 pounds of salt per gallon or 18.5 pounds per cubic foot.

- Refrigeration tonnage is gallons per minute of cooling water times increased temperature divided by 24.

- Cooling tower makeup is estimated at 1.5 gallons per hour per ton of refrigeration.

- 1 ton of refrigeration is 288,000 Btu.

Units of Measure and Conversions

Chemistry

The science of chemistry deals with the structure, composition, and changes in composition of matter, as well as with the laws that govern these changes. To understand and work successfully with the chemical phases of wastewater treatment such as coagulation, sedimentation, softening, disinfection, and chemical removal of various undesirable substances, a wastewater operator needs to know some basic chemistry concepts.

Key

Atomic Number
Symbol
Common Name
Atomic Mass (Weight)

Atomic weights conform to the 1961 values of the Commission on Atomic Weights.

Example:
6 C Carbon 12.01

Periodic Table of Elements

Period	1 IA	2 IIA	3 IIIB	4 IVB	5 VB	6 VIB	7 VIIB	8 VIIIB	9 VIIIB	10 VIIIB	11 IB	12 IIB	13 IIIA	14 IVA	15 VA	16 VIA	17 VIIA	18 VIIIA
1	1 H Hydrogen 1.00794																	2 He Helium 4.002602
2	3 Li Lithium 6.941	4 Be Beryllium 9.012182											5 B Boron 10.811	6 C Carbon 12.0107	7 N Nitrogen 14.00674	8 O Oxygen 15.9994	9 F Fluorine 18.998402	10 Ne Neon 20.1797
3	11 Na Sodium 22.98977	12 Mg Magnesium 24.3050											13 Al Aluminum 26.981538	14 Si Silicon 28.0855	15 P Phosphorus 30.973761	16 S Sulfur 32.066	17 Cl Chlorine 35.4527	18 Ar Argon 39.948
4	19 K Potassium 39.0983	20 Ca Calcium 40.078	21 Sc Scandium 44.955910	22 Ti Titanium 47.867	23 V Vanadium 50.9415	24 Cr Chromium 51.9961	25 Mn Manganese 54.938049	26 Fe Iron 55.8457	27 Co Cobalt 58.933200	28 Ni Nickel 58.6934	29 Cu Copper 63.546	30 Zn Zinc 65.39	31 Ga Gallium 69.723	32 Ge Germanium 72.61	33 As Arsenic 74.92160	34 Se Selenium 78.96	35 Br Bromine 79.904	36 Kr Krypton 83.80
5	37 Rb Rubidium 85.4678	38 Sr Strontium 87.62	39 Y Yttrium 88.90585	40 Zr Zirconium 91.224	41 Nb Niobium 92.90638	42 Mo Molybdenum 95.94	43 Tc Technetium (98)	44 Ru Ruthenium 101.07	45 Rh Rhodium 102.90550	46 Pd Palladium 106.42	47 Ag Silver 107.8682	48 Cd Cadmium 112.411	49 In Indium 114.818	50 Sn Tin 118.710	51 Sb Antimony 121.760	52 Te Tellurium 127.60	53 I Iodine 126.90447	54 Xe Xenon 131.29
6	55 Cs Caesium 132.90545	56 Ba Barium 137.327	57–71	72 Hf Hafnium 178.49	73 Ta Tantalum 180.9479	74 W Tungsten 183.84	75 Re Rhenium 186.207	76 Os Osmium 190.23	77 Ir Iridium 192.217	78 Pt Platinum 195.078	79 Au Gold 196.96655	80 Hg Mercury 200.59	81 Tl Thallium 204.3833	82 Pb Lead 207.2	83 Bi Bismuth 208.98038	84 Po Polonium (209)	85 At Astatine (210)	86 Rn Radon (222)
7	87 Fr Francium (223)	88 Ra Radium (226)	89–103	104 Rf Rutherfordium (261)	105 Db Dubnium (262)	106 Sg Seaborgium (263)	107 Bh Bohrium (262)	108 Hs Hassium (265)	109 Mt Meitnerium (266)	110 Uun Ununnilium (269)	111 Uuu Unununium (272)	112 Uub Ununbium (277)	113 Uut	114 Uuq Ununquadium (289)	115 Uup Ununpentium (289)	116 Uuh Ununhexium (289)	117 Uus	118 Uuo Ununoctium (293)

57 La Lanthanum 138.9055	58 Ce Cerium 140.116	59 Pr Praseodymium 140.90765	60 Nd Neodymium 114.24	61 Pm Promethium (145)	62 Sm Samarium 150.36	63 Eu Europium 151.964	64 Gd Gadolinium 157.25	65 Tb Terbium 158.92534	66 Dy Dysprosium 162.50	67 Ho Holmium 164.93032	68 Er Erbium 167.26	69 Tm Thulium 168.93421	70 Yb Ytterbium 173.04	71 Lu Lutetium 174.967
89 Ac Actinium (227)	90 Th Thorium 232.0381	91 Pa Protactinium 231.03588	92 U Uranium 238.0289	93 Np Neptunium (237)	94 Pu Plutonium (244)	95 Am Americium (243)	96 Cm Curium (247)	97 Bk Berkelium (247)	98 Cf Californium (251)	99 Es Einsteinium (252)	100 Fm Fermium (257)	101 Md Mendelevium (258)	102 No Nobelium (259)	103 Lr Lawrencium (262)

Atomic masses in parentheses are those of the most stable or common isotope.

Note: The subgroup numbers 1–18 were adopted in 1984 by the International Union of Pure and Applied Chemistry. The names of elements 110–118 are the Latin equivalents of those numbers.

List of Elements

Name	Symbol	Atomic Number	Atomic Weight
Actinium	Ac	89	227*
Aluminum	Al	13	26.98
Americium	Am	95	243*
Antimony	Sb	51	121.75
Argon	Ar	18	39.95
Arsenic	As	33	74.92
Astatine	At	85	210*
Barium	Ba	56	137.34
Berkelium	Bk	97	247*
Beryllium	Be	4	9.01
Bismuth	Bi	83	208.98
Boron	B	5	10.81
Bromine	Br	35	79.90
Cadmium	Cd	48	112.41
Calcium	Ca	20	40.08
Californium	Cf	98	251*
Carbon	C	6	12.01
Cerium	Ce	58	140.12
Cesium	Cs	55	132.91
Chlorine	Cl	17	35.45
Chromium	Cr	24	52.00
Cobalt	Co	27	58.93
Copper	Cu	29	63.55
Curium	Cm	96	247*
Dubnium	Db	105	262*
Dysprosium	Dy	66	162.50
Einsteinium	Es	99	252*
Erbium	Er	68	167.26
Europium	Eu	63	151.96
Fermium	Fm	100	257*
Fluorine	F	9	19.00

Chemistry

Table continued on next page

List of Elements (continued)

Name	Symbol	Atomic Number	Atomic Weight
Francium	Fr	87	223[*]
Gadolinium	Gd	64	157.25
Gallium	Ga	31	69.72
Germanium	Ge	32	72.64
Gold	Au	79	196.97
Hafnium	Hf	72	178.49
Hassium	Hs	108	265[*]
Helium	He	2	4.00
Holmium	Ho	67	164.93
Hydrogen	H	1	1.01
Indium	In	49	114.82
Iodine	I	53	126.90
Iridium	Ir	77	192.22
Iron	Fe	26	55.85
Krypton	Kr	36	83.80
Lanthanum	La	57	138.91
Lawrencium	Lr	103	262[*]
Lead	Pb	82	207.2
Lithium	Li	3	6.94
Lutetium	Lu	71	174.97
Magnesium	Mg	12	24.31
Manganese	Mn	25	54.94
Meitnerium	Mt	109	265[*]
Mendelevium	Md	101	258[*]
Mercury	Hg	80	200.59
Molybdenum	Mo	42	95.94
Neodymium	Nd	60	144.24
Neon	Ne	10	20.18
Neptunium	Np	93	237.05[†]
Nickel	Ni	28	58.69

Table continued on next page

List of Elements (continued)

Name	Symbol	Atomic Number	Atomic Weight
Niobium	Nb	41	92.91
Nitrogen	N	7	14.01
Nobelium	No	102	259[*]
Osmium	Os	76	190.23
Oxygen	O	8	16.00
Palladium	Pd	46	106.42
Phosphorus	P	15	30.97
Platinum	Pt	78	195.08
Plutonium	Pu	94	244[*]
Polonium	Po	84	209[*]
Potassium	K	19	39.10
Praseodymium	Pr	59	140.91
Promethium	Pm	61	145[*]
Protactinium	Pa	91	231.04[†]
Radium	Ra	88	226.03[†]
Radon	Rn	86	222[*]
Rhenium	Re	75	186.21
Rhodium	Rh	45	102.91
Rubidium	Rb	37	85.47
Ruthenium	Ru	44	101.07
Rutherfordium	Rf	104	261[*]
Samarium	Sm	62	150.36
Scandium	Sc	21	44.96
Seaborgium	Sg	106	263[*]
Selenium	Se	34	78.96
Silicon	Si	14	28.09
Silver	Ag	47	107.87
Sodium	Na	11	22.99
Strontium	Sr	38	87.62
Sulfur	S	16	32.06

Table continued on next page

Chemistry

List of Elements (continued)

Name	Symbol	Atomic Number	Atomic Weight
Tantalum	Ta	73	180.95
Technetium	Tc	43	98.91[†]
Tellurium	Te	52	127.60
Terbium	Tb	65	158.93
Thallium	Tl	81	204.38
Thorium	Th	90	232.04[†]
Thulium	Tm	69	168.93
Tin	Sn	50	118.69
Titanium	Ti	22	47.90
Tungsten	W	74	183.85
Ununbium	Uub	112	27[*]
Ununnillium	Uun	110	269[*]
Ununhexium	Uuh	116	289[*]
Ununoctium	Uuo	118	293[*]
Ununquadium	Uuq	114	285[*]
Unununium	Uuu	111	27[*]
Uranium	U	92	238.03
Vanadium	V	23	50.94
Xenon	Xe	54	131.29
Ytterbium	Yb	70	173.04
Yttrium	Y	39	88.91
Zinc	Zn	30	65.38
Zirconium	Zr	40	91.22

* Mass number of most stable or best-known isotope.
† Mass of most commonly available, long-lived isotope.

Compounds Common in Wastewater Treatment

Chemical Name	Common Name	Chemical Formula
Aluminum hydroxide	Alum floc	$Al(OH)_3$
Aluminum sulfate	Filter alum	$Al_2(SO_4)_3 \cdot 14H_2O$
Ammonia	Ammonia	NH_3 (ammonia gas)
Calcium bicarbonate	—	$Ca(HCO_3)_2$
Calcium carbonate	Limestone	$CaCO_3$
Calcium chloride	—	$CaCl_2$
Calcium hydroxide	Hydrated lime (slaked lime)	$Ca(OH)_2$
Calcium hypochlorite	HTH	$Ca(OCl)_2$
Calcium oxide	Unslaked lime (quicklime)	CaO
Calcium sulfate	—	$CaSO_4$
Carbon	Activated carbon	C
Carbon dioxide	—	CO_2
Carbonic acid	—	H_2CO_3
Chlorine	—	Cl_2
Chlorine dioxide	—	ClO_2
Copper sulfate	Blue vitriol	$CuSO_4 \cdot 5H_2O$
Dichloramine	—	$NHCl_2$
Ferric chloride	—	$FeCl_3 \cdot 6H_2O$
Ferric hydroxide	Ferric hydroxide floc	$Fe(OH)_3$
Ferric sulfate	—	$Fe_2(SO_4)_3 \cdot 3H_2O$
Ferrous bicarbonate	—	$Fe(HCO_3)_2$
Ferrous hydroxide	—	$Fe(OH)_2$
Fluosilicic acid (hydrofluosilicic acid)	—	H_2SiF_6
Hydrochloric acid	Muriatic acid	HCl
Hydrofluosilicic acid (fluosilicic acid)	—	H_2SiF_6
Hydrogen sulfide	—	H_2S
Hypochlorous acid	—	$HOCl$
Magnesium bicarbonate	—	$Mg(HCO_3)_2$
Magnesium carbonate	—	$MgCO_3$
Magnesium chloride	—	$MgCl_2$
Magnesium hydroxide	—	$Mg(OH)_2$

Chemistry

Table continued on next page

Compounds Common in Wastewater Treatment (continued)

Chemical Name	Common Name	Chemical Formula
Manganese dioxide	—	MnO_2
Manganous bicarbonate	—	$Mn(HCO_3)_2$
Manganous sulfate	—	$MnSO_4$
Monochloramine	—	NH_2Cl
Potassium bicarbonate	—	$KHCO_3$
Potassium permanganate	—	$KMnO_4$
Sodium bicarbonate	Soda	$NaHCO_3$
Sodium carbonate	Soda ash	Na_2CO_3
Nitrogen trichloride (trichloramine)	—	NCl_3
Sodium chloride	Salt	$NaCl$
Sodium chlorite	—	$NaClO_2$
Sodium fluoride	—	NaF
Sodium fluosilicate (sodium silicofluoride)	—	Na_2SiF_6
Sodium hydroxide	Lye	$NaOH$
Sodium hypochlorite	Bleach	$NaOCl$
Sodium phosphate	—	$Na_3PO_4 \cdot 12H_2O$
Sodium silicofluoride (sodium fluosilicate)	—	Na_2SiF_6
Sodium bisulfite	—	$NaHSO_3$
Sodium sulfate	—	Na_2SO_4
Sodium sulfite	—	Na_2SO_3
Sodium thiosulfate	—	$Na_2S_2O_3 \cdot 5H_2O$
Sulfur dioxide	—	SO_2
Sulfuric acid	Oil of vitriol	H_2SO_4
Trichloramine (nitrogen trichloride)	—	NCl_3

KEY FORMULAS FOR CHEMISTRY _____

total suspended solids, mg/L =

$$\frac{\text{paper wt. and dried solids, g} - \text{paper wt., g}}{\text{mL of sample}} \times 1{,}000{,}000$$

total solids, mg/L $= \dfrac{\text{residue, mg} \times 1{,}000}{\text{mL sample}}$

total alkalinity, mg/L $= \dfrac{\text{mL of titrant} \times \text{normality} \times 50{,}000}{\text{mL of sample}}$

Langelier saturation index = pH − pH, saturated

Concentrations

concentration 1 × volume 1 = concentration 2 × volume 2

$\text{mg/L} = \dfrac{\text{mL} \times 1{,}000{,}000}{\text{mL sample}} = \text{mL} \times 1{,}000 \text{ mL/L}$

$\text{mg/L total solids} = \dfrac{\text{residue, mg} \times 1{,}000}{\text{mL sample}}$

$\text{percent strength by weight} = \dfrac{\text{weight of solute}}{\text{weight of solution}} \times 100$

$\text{number of moles} = \dfrac{\text{total weight}}{\text{molecular weight}}$

$\text{molarity} = \dfrac{\text{moles of solute}}{\text{liters of solution}}$

$\text{number of equivalent weights} = \dfrac{\text{total weight}}{\text{equivalent weight}}$

$\text{normality} = \dfrac{\text{number of equivalent weights of solute}}{\text{liters of solution}}$

$\left(\dfrac{\text{molecular weight of new measure}}{\text{molecular weight of old measure}} \right) (\text{old concentration}) = \text{new concentration}$

Chemistry

High Concentration of H^+ Ions	H^+ and OH^- Ions in Balance	High Concentration of OH^- Ions
0 — 1 — 2 — 3 — 4 — 5 — 6 — 7 — 8 — 9 — 10 — 11 — 12 — 13 — 14		
Pure Acid	Neutral	Pure Base

The pH Scale

CONDUCTIVITY AND DISSOLVED SOLIDS

Electrical conductivity is the ability of a solution to conduct an electric current and it can be used as an indirect measure of the total dissolved solids (TDS) in a water sample.

The unit of measure commonly used is siemens per centimeter (S/cm). The conductivity of water is usually expressed as microsiemens per centimeter (μS/cm) which is 10^{-6} S/cm. The relationship between conductivity and dissolved solids is approximately:

$$2 \; \mu S/cm = 1 \; ppm \text{ (which is the same as 1 mg/L)}$$

The conductivity of water from various sources is

Absolutely pure water	= 0.055 μS/cm
Distilled water	= 0.5 μS/cm
Mountain water	= 1.0 μS/cm
Most drinking water sources	= 500 to 800 μS/cm
Seawater	= 56 mS/cm
Maximum for potable water	= 1,055 μS/cm

Some common conductivity conversion factors are

mS/cm	×	1,000	=	μS/cm
μS/cm	×	0.001	=	mS/cm
μS/cm	×	1	=	μmhos/cm
μS/cm	×	0.5	=	mg/L of TDS
mS/cm	×	0.5	=	g/L of TDS
mg/L TDS	×	0.001	=	g/L of TDS
mg/L TDS	×	0.05842	=	gpg TDS

Densities of Various Substances

Substance	Density lb/ft³	lb/gal
Solids		
Activated carbon[*†]	8–28 (avg. 12)	
Lime[*†]	20–50	
Dry alum[*†]	60–75	
Aluminum (at 20°C)	168.5	
Steel (at 20°C)	486.7	
Copper (at 20°C)	555.4	
Liquids		
Propane (−44.5°C)	36.5	4.88
Gasoline[†]	43.7	5.84
Water (4°C)	62.4	8.34
Fluorosilicic acid (30%, −8.1°C)	77.8–79.2	10.4–10.6
Liquid alum (36°Bé, 15.6°C)	83.0	11.09
Liquid chlorine (−33.6°C)	97.3	13.01
Sulfuric acid (18°C)	114.2	15.3
Gases		
Methane (0°C, 14.7 psia)	0.0344	
Air (20°C, 14.7 psia)	0.075	
Oxygen (0°C, 14.7 psia)	0.089	
Hydrogen sulfide[†]	0.089	
Carbon dioxide[†]	0.115	
Chlorine gas (0°C, 14.7 psia)	0.187	

[*] Bulk density of substance.
[†] Temperature and/or pressure not given.

The density of granite rock is about 162 lb/ft³, and the density of water is 62.4 lb/ft³. The specific gravity of granite is found by this ratio:

$$\text{specific gravity} = \frac{\text{density of granite}}{\text{density of water}} = \frac{162 \text{ lb}/\text{ft}^3}{62.4 \text{ lb}/\text{ft}^3} = 2.60$$

Chemistry

Specific Gravities of Various Solids and Liquids

Substance	Specific Gravity
Solids	
Aluminum (20°C)	2.7
Steel (20°C)	7.8
Copper (20°C)	8.9
Activated carbon[*†]	0.13–0.45 (avg. 0.19)
Lime[*†]	0.32–0.80
Dry alum[*†]	0.96–1.2
Soda ash[*†]	0.48–1.04
Coagulant aids (polyelectrolytes)[*†]	0.43–0.56
Table salt[*†]	0.77–1.12
Liquids	
Liquid alum (36°Bé, 15.6°C)	1.33
Water (4°C)	1.00
Fluorosilicic acid (30%, −8.1°C)	1.25–1.27
Sulfuric acid (18°C)	1.83
Ferric chloride (30%, 30°C)	1.34

*Bulk density used to determine specific gravity.
†Temperature and/or pressure not given.

Specific Gravities of Various Gases

Gas		Specific Gravity
Hydrogen (0°C; 14.7 psia)	0.07	When released in a room, these
Methane (0°C; 14.7 psia)	0.46	gases will first rise to the ceiling
Carbon monoxide[*]	0.97	area.
Air (20°C; 14.7 psia)	1.00	
Nitrogen (0°C; 14.7 psia)	1.04	When released in a room, these
Oxygen (0°C; 14.7 psia)	1.19	gases will first settle to the floor
Hydrogen sulfide*	1.19	area.
Carbon dioxide*	1.53	
Chlorine gas (0°C; 14.7 psia)	2.49	
Gasoline vapor*	3.0	

*Temperature and pressure not given.

Oxidation Numbers of Various Elements

Element	Common Valences	Element	Common Valences
Arsenic (As)	+3, +5	Magnesium (Mg)	+2
Barium (Ba)	+2	Mercury (Hg)	+1, +2
Cadmium (Cd)	+2	Nitrogen (N)	+3, −3, +5
Calcium (Ca)	+2	Oxygen (O)	−2
Carbon (C)	+4, −4	Phosphorus (P)	−3
Chlorine (Cl)	−1	Potassium (K)	+1
Chromium (Cr)	+3	Selenium (Se)	−2, +4
Copper (Cu)	+1, +2	Sulfur (S)	−2, +4, +6
Hydrogen (H)	+1		

Oxidation Numbers of Common Radicals

Radical	Common Valences
Ammonium (NH_4)	+1
Bicarbonate (HCO_3)	−1
Hydroxide (OH)	−1
Nitrate (NO_3)	−1
Nitrite (NO_2)	−1
Carbonate (CO_3)	−2
Sulfate (SO_4)	−2
Sulfite (SO_3)	−2
Phosphate (PO_4)	−3

Chemistry

Dilution

$$\left(\begin{array}{c}\text{normality of}\\\text{solution 1}\end{array}\right)\left(\begin{array}{c}\text{volume of}\\\text{solution 1}\end{array}\right) = \left(\begin{array}{c}\text{normality of}\\\text{solution 2}\end{array}\right)\left(\begin{array}{c}\text{volume of}\\\text{solution 2}\end{array}\right)$$

This equation can be abbreviated as

$$(N_1)(V_1) = (N_2)(V_2)$$

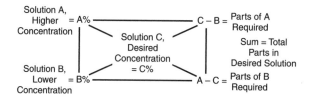

Solution A, Higher Concentration = A%

Solution C, Desired Concentration = C%

Solution B, Lower Concentration = B%

C − B = Parts of A Required

Sum = Total Parts in Desired Solution

A − C = Parts of B Required

Rectangle Method (sometimes called the dilution rule)

Some Chemicals Used in Water and Wastewater Treatment

Chemical Name	Common Name	Chemical Formula	Used for
Aluminum oxide	Alumina	Al_2O_3	Fluoride, arsenic removal
Aluminum sulfate	Alum	$Al_2(SO_4)_3 \cdot 14H_2O$	Coagulation
Ammonia	Ammonia gas	NH_3 (ammonia gas)	
	Ammonia aqua	NH_4OH (ammonia solution)	Chloramination
Calcium bicarbonate		$Ca(HCO_3)_2$	Alkalinity
Calcium carbonate	Limestone	$CaCO_3$	
Calcium hydroxide	Hydrated lime or slaked lime	$Ca(OH)_2$	Softening
Calcium hypochlorite	HTH	$Ca(ClO)_2$	Chlorination
Calcium oxide	Unslaked lime or quick lime	CaO	Softening
Carbon	Activated carbon	C	Taste, odors, and pesticide removal
Chlorine		Cl_2	Disinfection
Chlorine dioxide		ClO_2	Disinfection
Copper sulfate	Blue vitriol	$CuSO_4 \cdot 5H_2O$	Algae control
Ferric chloride		$FeCl_3 \cdot 6H_2O$	Coagulation
Ferric sulfate		$Fe_2(SO_4)_3$	Coagulation
Ferrous chloride		$FeCl_2$	Chlorite control
Fluosilicic acid (hydrofluosilicic acid)	Fluoride	H_2SiF_6	Fluoridation
Hydrochloric acid	Muriatic acid	HCl	
Ozone		O_3	Disinfection
Potassium dichromate		$K_2Cr_2O_7$	
Potassium permanganate		$KMnO_4$	Taste and odor control
Sodium aluminate		$NaAlO_2$	Coagulation
Sodium bicarbonate	Baking soda	$NaHCO_3$	Alkalinity
Sodium carbonate	Soda ash	Na_2CO_3	Softening
Sodium chloride	Salt	$NaCl$	
Sodium chlorite		$NaClO_2$	Chlorine dioxide formation
Sodium fluoride		NaF	Fluoridation
Sodium fluosilicate		Na_2SiF_6	Fluoridation
Sodium hexametaphosphate	Calgon	$Na_6(PO_3)_6$ or $6(NaPO_3)$	Sequestering
Sodium hydroxide	Lye	$NaOH$	Alkalinity
Sodium hypochlorite	Bleach	$NaClO$	Chlorination
Sodium phosphate		$Na_3PO_4 \cdot 12H_2O$	
Sodium thiosulfate		$Na_2S_2O_3$	
Sulfuric acid	Oil of vitriol	H_2SO_4	
Zinc orthophosphate		$Zn_3(PO_4)_2$	Corrosion control

Nitrification Reaction

Biological nitrification is an aerobic autotrophic process in which the energy for bacterial growth is derived from the oxidation of inorganic compounds, primarily ammonia nitrogen. Autotrophic nitrifiers, in contrast to heterotrophs, use inorganic carbon dioxide instead of organic carbon for cell synthesis. The yield of nitrifier cells per unit of substrate metabolized is many times smaller than that for heterotrophic bacteria.

Although a variety of nitrifying bacteria exist in nature, the two genera associated with biological nitrification are *Nitrosomonas* and *Nitrobacter*. The oxidation of ammonia to nitrate is a two-step process requiring both nitrifiers for the conversion. *Nitrosomonas* oxidizes ammonia to nitrite, while *Nitrobacter* subsequently transforms nitrite to nitrate. The respective oxidation reactions are as follows:

Ammonia oxidation:

$$NH_4^+ + 1.5O_2 + 2HCO_3^- \xrightarrow{\text{\textit{Nitrobacter}}} NO_2^- + 2H_2CO_3 + H_2O$$

Nitrite oxidation:

$$NO_2^- + 0.5O_2 \xrightarrow{\text{\textit{Nitrobacter}}} NO_3^-$$

Overall reaction:

$$NH_4^+ + 2O_2 + 2HCO_3^- \xrightarrow{\text{nitrifiers}} NO_3^- + 2H_2CO_3 + H_2O$$

Dissolved-Oxygen Concentration in Water as a Function of Temperature and Salinity (barometric pressure = 760 mm Hg)

Temperature, °C	Dissolved-Oxygen Concentration, mg/L									
	Salinity, ppt [*]									
	0	5	10	15	20	25	30	35	40	45
0	14.60	14.11	13.64	13.18	12.74	12.31	11.90	11.50	11.11	10.74
1	14.20	13.73	13.27	12.83	12.40	11.98	11.58	11.20	10.83	10.46
2	13.81	13.36	12.91	12.49	12.07	11.67	11.29	10.91	10.55	10.20
3	13.45	13.00	12.58	12.16	11.76	11.38	11.00	10.64	10.29	9.95
4	13.09	12.67	12.25	11.85	11.47	11.09	10.73	10.38	10.04	9.71
5	12.76	12.34	11.94	11.56	11.18	10.82	10.47	10.13	9.80	9.48
6	12.44	12.04	11.65	11.27	10.91	10.56	10.22	9.89	9.57	9.27
7	12.13	11.74	11.37	11.00	10.65	10.31	9.98	9.66	9.35	9.06
8	11.83	11.46	11.09	10.74	10.40	10.07	9.75	9.44	9.14	8.85
9	11.55	11.19	10.83	10.49	10.16	9.84	9.53	9.23	8.94	8.66
10	11.28	10.92	10.58	10.25	9.93	9.62	9.32	9.03	8.75	8.47
11	11.02	10.67	10.34	10.02	9.71	9.41	9.12	8.83	8.56	8.30
12	10.77	10.43	10.11	9.80	9.50	9.21	8.92	8.65	8.38	8.12
13	10.53	10.20	9.89	9.59	9.30	9.01	8.74	8.47	8.21	7.96
14	10.29	9.98	9.68	9.38	9.10	8.82	8.55	8.30	8.04	7.80
15	10.07	9.77	9.47	9.19	8.91	8.64	8.38	8.13	7.88	7.65
16	9.86	9.56	9.28	9.00	8.73	8.47	8.21	7.97	7.73	7.50
17	9.65	9.36	9.09	8.82	8.55	8.30	8.05	7.81	7.58	7.36
18	9.45	9.17	8.90	8.64	8.39	8.14	7.90	7.66	7.44	7.22
19	9.26	8.99	8.73	8.47	8.22	7.98	7.75	7.52	7.30	7.09
20	9.08	8.81	8.56	8.31	8.07	7.83	7.60	7.38	7.17	6.96

Table continued on next page

Dissolved-Oxygen Concentration in Water as a Function of Temperature and Salinity (barometric pressure = 760 mm Hg) (continued)

Temperature, °C	Dissolved-Oxygen Concentration, mg/L									
	Salinity, ppt*									
	0	5	10	15	20	25	30	35	40	45
21	8.90	8.64	8.39	8.15	7.91	7.69	7.46	7.25	7.04	6.84
22	8.73	8.48	8.23	8.00	7.77	7.54	7.33	7.12	6.91	6.72
23	8.56	8.32	8.08	7.85	7.63	7.41	7.20	6.99	6.79	6.60
24	8.40	8.16	7.93	7.71	7.49	7.28	7.07	6.87	6.68	6.49
25	8.24	8.01	7.79	7.57	7.36	7.15	6.95	6.75	6.56	6.38
26	8.09	7.87	7.65	7.44	7.23	7.03	6.83	6.64	6.46	6.28
27	7.95	7.73	7.51	7.31	7.10	6.91	6.72	6.53	6.35	6.17
28	7.81	7.59	7.38	7.18	6.98	6.79	6.61	6.42	6.25	6.08
29	7.67	7.46	7.26	7.06	6.87	6.68	6.50	6.32	6.15	5.98
30	7.54	7.33	7.14	6.94	6.75	6.57	6.39	6.22	6.05	5.89
31	7.41	7.21	7.02	6.83	6.65	6.47	6.29	6.12	5.96	5.80
32	7.29	7.09	6.90	6.72`	6.54	6.36	6.19	6.03	5.87	5.71
33	7.17	6.98	6.79	6.61	6.44	6.26	6.10	5.94	5.78	5.63
34	7.05	6.86	6.68	6.51	6.33	6.17	6.01	5.85	5.69	5.54
35	6.93	6.75	6.58	6.40	6.24	6.07	5.92	5.76	5.61	5.46
36	6.82	6.65	6.47	6.31	6.14	5.98	5.83	5.68	5.53	5.39
37	6.72	6.54	6.37	6.21	6.05	5.89	5.74	5.59	5.45	5.31
38	6.61	6.44	6.28	6.12	5.96	5.81	5.66	5.51	5.37	5.24
39	6.51	6.34	6.18	6.03	5.87	5.72	5.58	5.44	5.30	5.16
40	6.41	6.25	6.09	5.94	5.79	5.64	5.50	5.36	5.22	5.09

* ppt = parts per thousand.

Chemistry

Dissolved-Oxygen Concentration in Water as a Function of Temperature and Barometric Pressure (salinity = 0 ppt[*])

Temperature, °C	Dissolved-Oxygen Concentration, mg/L									
	Barometric Pressure, mm of mercury									
	735	740	745	750	755	760	765	770	775	780
0	14.12	14.22	14.31	14.41	14.51	14.60	14.70	14.80	14.89	14.99
1	13.73	13.82	13.92	14.01	14.10	14.20	14.29	14.39	14.48	14.57
2	13.36	13.45	13.54	13.63	13.72	13.81	13.90	14.00	14.09	14.18
3	13.00	11.09	13.18	13.27	13.36	11.45	13.53	13.62	13.71	13.80
4	12.66	12.75	12.83	12.92	13.01	13.09	13.18	13.27	13.35	13.44
5	12.33	12.42	12.50	12.59	12.67	12.76	12.84	12.93	13.01	13.10
6	12.02	12.11	12.19	12.27	12.35	12.44	12.52	12.60	12.68	12.77
7	11.72	11.80	11.89	11.97	12.05	12.13	12.21	12.29	12.37	12.45
8	11.44	11.52	11.60	11.67	11.75	11.83	11.91	11.99	12.07	12.15
9	11.16	11.24	11.32	11.40	11.47	11.55	11.63	11.70	11.78	11.86
10	10.90	10.98	11.05	11.13	11.20	11.28	11.35	11.43	11.50	11.58
11	10.65	10.72	10.80	10.87	10.94	11.02	11.09	11.16	11.24	11.31
12	10.41	10.48	10.55	10.62	10.69	10.77	10.84	10.91	10.98	11.05
13	10.17	10.24	10.31	10.38	10.46	10.53	10.60	10.67	10.74	10.81
14	9.95	10.02	10.09	10.16	10.23	10.29	10.36	10.43	10.50	10.57
15	9.73	9.80	9.87	9.94	10.00	10.07	10.14	10.21	10.27	10.34
16	9.53	9.59	9.66	9.73	9.79	9.86	9.92	9.99	10.06	10.12
17	9.33	9.39	9.46	9.52	9.59	9.65	9.72	9.78	9.85	9.91
18	9.14	9.20	9.26	9.33	9.39	9.45	9.52	9.58	9.64	9.71
19	8.95	9.01	9.07	9.14	9.20	9.26	9.32	9.39	9.45	9.51
20	8.77	8.83	8.89	8.95	9.02	9.08	9.14	9.20	9.26	9.32

Table continued on next page

Dissolved-Oxygen Concentration in Water as a Function of Temperature and Barometric Pressure (salinity = 0 ppt[*]) (continued)

Temperature, °C	Dissolved-Oxygen Concentration, mg/L									
	Barometric Pressure, mm of mercury									
	735	740	745	750	755	760	765	770	775	780
21	8.60	8.66	8.72	8.78	8.84	8.90	8.96	9.02	9.08	9.14
22	8.43	8.49	8.55	8.61	8.67	8.73	8.79	8.84	8.90	8.96
23	8.27	8.33	8.39	8.44	8.50	8.56	8.62	8.68	8.73	8.79
24	8.11	8.17	8.23	8.29	8.34	8.40	8.46	8.51	8.57	8.63
25	7.96	8.02	8.08	8.13	8.19	8.24	8.30	8.36	8.41	8.47
26	7.82	7.87	7.93	7.98	8.04	8.09	8.15	8.20	8.26	8.31
27	7.68	7.73	7.79	7.84	7.89	7.95	8.00	8.06	8.11	8.17
28	7.54	7.59	7.65	7.70	7.75	7.81	7.86	7.91	7.97	8.02
29	7.41	7.46	7.51	7.57	7.62	7.67	7.72	7.78	7.83	7.88
30	7.28	7.33	7.38	7.44	7.49	7.54	7.59	7.64	7.69	7.75
31	7.16	7.21	7.26	7.31	7.36	7.41	7.46	7.51	7.46	7.62
32	7.04	7.09	7.14	7.19	7.24	7.29	7.34	7.39	7.44	7.49
33	6.92	6.97	7.02	7.07	7.12	7.17	7.22	7.27	7.31	7.36
34	6.80	6.85	6.90	6.95	7.00	7.05	7.10	7.15	7.20	7.24
35	6.69	6.74	6.79	6.84	6.89	6.93	6.98	7.03	7.08	7.13
36	6.59	6.63	6.68	6.73	6.78	6.82	6.87	6.92	6.97	7.01
37	6.48	6.53	6.57	6.62	6.67	6.72	6.76	6.81	6.86	6.90
38	6.38	6.43	6.47	6.52	6.56	6.61	6.66	6.70	6.75	6.80
39	6.28	6.33	6.37	6.42	6.46	6.51	6.56	6.60	6.65	6.69
40	6.18	6.23	6.27	6.32	6.36	6.41	6.46	6.50	6.55	6.59

* ppt = parts per thousand.

Chemistry

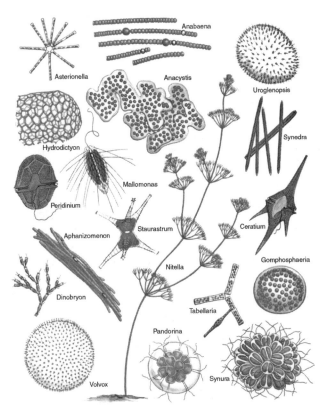

Source: Standard Methods for the Examination of Water and Wastewater.

Taste and Odor Algae

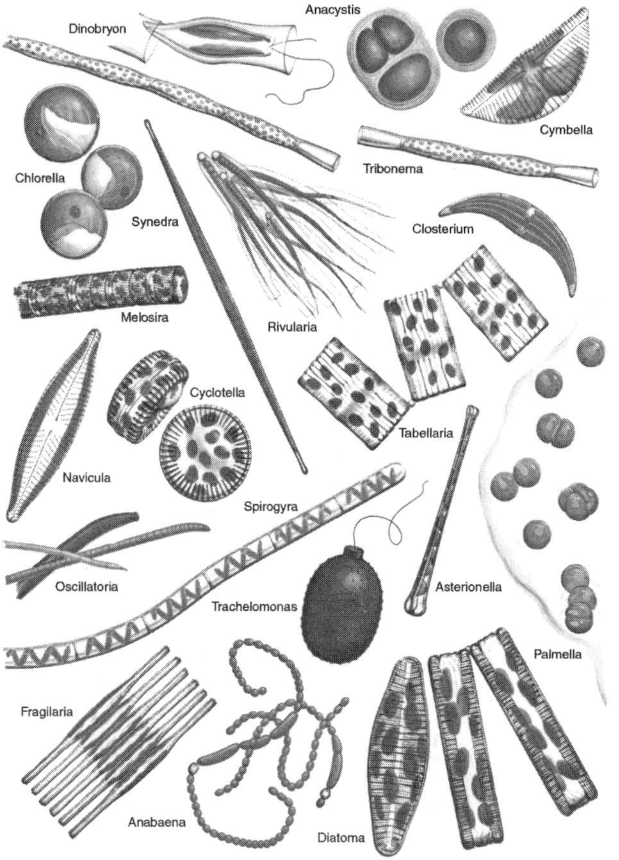
Anacystis
Dinobryon
Cymbella
Tribonema
Chlorella
Synedra
Closterium
Melosira
Rivularia
Cyclotella
Tabellaria
Navicula
Spirogyra
Oscillatoria
Asterionella
Trachelomonas
Palmella
Fragilaria
Anabaena
Diatoma

Source: Standard Methods for the Examination of Water and Wastewater.

Filter- and Screen-Clogging Algae

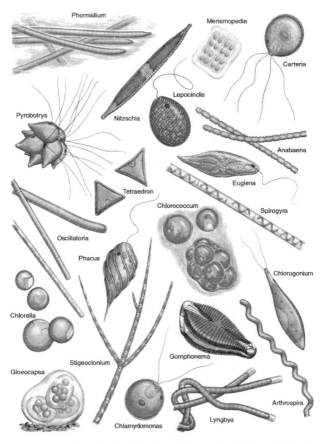

Source: Standard Methods for the Examination of Water and Wastewater.

Freshwater Pollution Algae

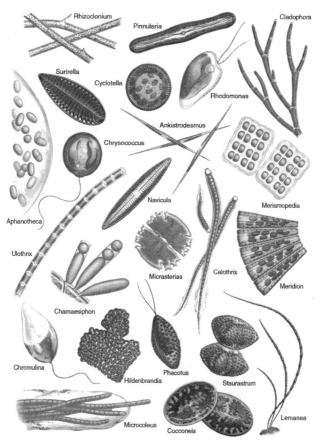

Source: Standard Methods for the Examination of Water and Wastewater.

Clean Water Algae

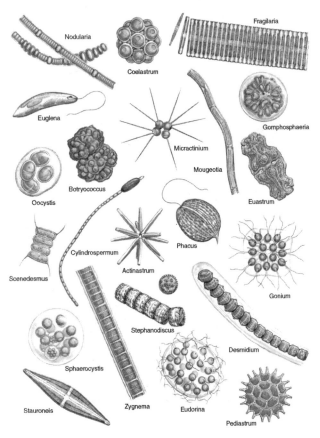

Source: Standard Methods for the Examination of Water and Wastewater.

Plankton and Other Surface Water Algae

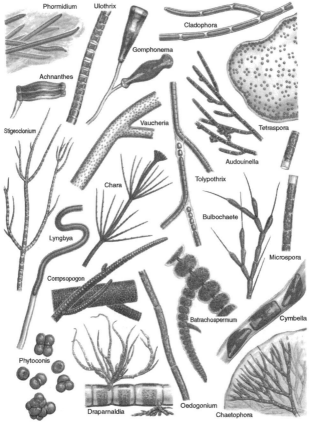

Source: Standard Methods for the Examination of Water and Wastewater.

Algae Growing on Surfaces

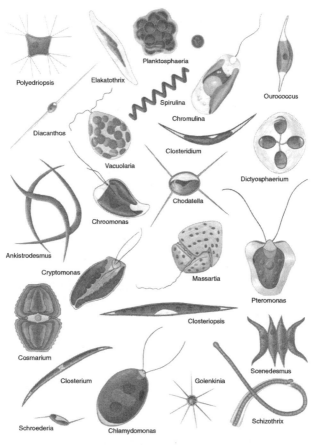

Source: Standard Methods for the Examination of Water and Wastewater.

Wastewater Treatment Pond Algae

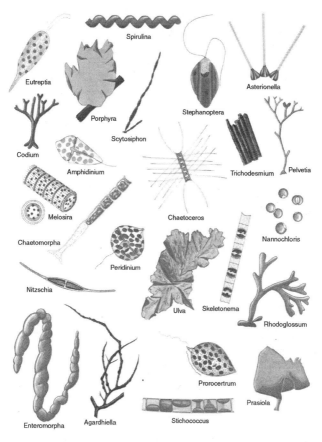

Source: Standard Methods for the Examination of Water and Wastewater.

Estuarine Pollution Algae

Safety

Wastewater operators are exposed to a number of occupational hazards. In fact, water and wastewater treatment ranks high on the national listings of industrial occupations where on-the-job injuries can occur. Whether regulated by the Occupational Safety and Health Administration or dictated by common sense and plant policy, safe working practices are an important part of the wastewater operator's job.

Pipeline Color Coding Used in Wastewater Treatment Plants

Type of Line	Contents of Line	Color of Pipe
Sludge lines	Raw sludge	Brown with black bands
	Sludge recirculation or suction	Brown with yellow bands
	Sludge draw off	Brown with orange bands
	Sludge recirculation discharge	Brown
Gas lines	Sludge gas	Orange (or red)
	Natural gas	Orange (or red) with black bands
Water lines	Nonpotable water	Blue with black bands
	Potable water	Blue
	Water for heating digestors or buildings	Blue with 6-in. (150-mm) red bands spaced 30 in. (760 mm) apart
	Reuse	Purple
Other lines	Chlorine	Yellow
	Sulfur dioxide	Yellow with red bands
	Sewage (wastewater)	Gray
	Compressed air	Green

Source: Recommended Standards for Water Works and Recommended Standards for Wastewater Facilities *(the "Ten States Standards").*

NOTE: It is recommended that the direction of flow and name of the contents be noted on all lines.

Confined Space Entry

Beginning in April 1993, the Occupational Safety and Health Administration (OSHA) implemented and started enforcing comprehensive regulations governing confined spaces. Most states and municipalities have adopted these standards, even if OSHA does not regulate them directly.

Virtually all access entrances now come under OSHA standard 29 CFR 1910.146, *Permit-Required Confined Spaces*. These standards formally implement requirements and clarify previous recommendations and suggestions made by industry representatives.

Emergency Rescue

As of April 15, 1993, a mechanical device for rescue became required for all vertical-type, permit-required confined spaces deeper than 5 ft [1910.146(k)(3)(ii)]. A safety line and human muscles are no longer acceptable means of rescue for most confined spaces with the potential for vertical rescue. Systems that were used in the past, including "boat winches," should no longer be used. Today, "human-rated" alternatives are available that satisfy the OSHA requirements. This means that the manufacturer has designed the system specifically for lifting people rather than materials.

Nonemergency Ingress/Egress

Means for safe entry and exit by authorized personnel are just as important, per 1910.146(d)(4), as rescue systems. Most tripod/winch systems used for nonemergency work positioning and support applications (such as lowering a worker into an access space that does not contain a ladder) are defined as "single-point adjustable suspension scaffolds." Tripods and davit-arms are examples. Both general industry standards (OSHA 1910) and construction industry standards (OSHA 1926) stipulate specific requirements that must be satisfied when a tripod/manually operated winch

Safety

system is the primary means used to suspend or support workers. Excerpts from the standards follow.

Utility owners and operators are also now clearly responsible for contractor or subcontractor activities in and around confined spaces. Contractors should be trained in following proper procedures and using the right equipment.

I. a. OSHA 1910.28(i)(1) Single-point adjustable suspension scaffolds. The scaffolding [tripod, davit-arm], including power units or manually operated winches, shall be a type tested and listed by a nationally recognized testing laboratory.

b. OSHA 1926.451(k)(1) Single-point adjustable suspension scaffolds. The scaffolding [tripod, davit-arm], including power units or manually operated winches, shall be a type tested and listed by Underwriters Laboratories or Factory Mutual Engineering Corporation.

Confined Space Entry Procedure

Job: Manhole Inspection and Cleaning   Employee:_____   Date:_____

Dept:_____   Foreman:_____   Review Date:_____

Municipality:_____

Required and/or Recommended personal protection equipment (PPE): Coveralls, rubber gloves, safety boots, safety glasses, hard hat, immunizations.

Sequence of Basic Job Steps	Potential Accidents or Hazards	Safe Job Procedure
1. Secure the work site to ensure traffic and public safety.	Injury or damage to equipment by contact with vehicles. Injury to public, either pedestrians or vehicle occupants.	Follow traffic control plan.
2. Check manhole for hazardous gases that may be present and toxic vapors.	Ignition of gases that may be present and toxic vapors.	Follow procedures for confined space entry.
3. Remove access cover.	Injury to back or foot; slips and falls.	Always use proper access cover lifting tools.

Table continued on next page

Safety

Confined Space Entry Procedure (continued)

Sequence of Basic Job Steps	Potential Accidents or Hazards	Safe Job Procedure
4. Before entering confined space, use flashlight or mirror to visually check condition of manhole and ladder rungs. Ensure that testing of hazardous gases is continuous and ventilation is in use where entry is required.	Falls, hazardous gases, and infection.	Test the atmosphere of confined space for oxygen deficiency, explosive or toxic gas (confined space entry plan). Provide adequate lighting with explosion-proof fixture. Always wear hard hat. Wear rubber gloves. Wherever possible, carry out the job in such a manner so that entry of personnel into the manhole is not necessary. Ensure gas mask (self-contained breathing apparatus plan) and other safety equipment are operational and available. Ensure that life support and rescue equipment is available.
5. Perform routine flushing operation, removing debris and sediment as necessary.	Hazardous gases may be released from disturbed sediments. Surcharging of collection system. Slips, falls, and infection.	Wear hard hat at all times (PPE plan). Where entry into manhole is necessary, provide full body harness and lifeline and approved equipment for removing debris.
6. Replace access cover.	Injury to back or foot; slips and falls.	Use proper tools to clean the ring to allow the cover to fit snugly. Replace the cover and ensure that it fits properly.

Entry Date: _____ Start Time: _____ Completion Time: _____

Description of Work To Be Performed: _____

Description of Space

Confined Space ID Number: _____ Type: _____ Classification: _____

Building Name:

Location of Confined Space:

Entry Checklist

Potential Hazards Identified?	☐ Yes	☐ No
Communications Established With Operations Center?	☐ Yes	☐ No
Emergency Procedures Reviewed?	☐ Yes	☐ No
Entrants and Attendants Trained?	☐ Yes	☐ No
Isolation of Energy Completed?	☐ Yes	☐ No
Area Secured?	☐ Yes	☐ No
Emergency Escape Retrieval Equipment Available?	☐ Yes	☐ No
Personal Protective Equipment Used?	☐ Yes	☐ No

Confined Space Equipment and PPE Used During Entry

☐ Tripod With Mechanical Winch
☐ Rescue Tripod With Lifeline
☐ Harness
☐ Two-Way Communications
☐ General/Local Exhaust Ventilation
☐ Air Purifying Respirator
☐ Self-Contained Breathing Apparatus

☐ Steel Toe Boots
☐ Hard Hat
☐ Safety Glasses/Goggles/Face Shield
☐ Gloves
☐ Chemical Resistant Clothing
☐ Hearing Protection
Other PPE or Equipment Used:
_____
_____

Air Monitoring Results Prior to Entry

Monitor Type: _____ Serial Number: _____

Oxygen _____ % LEL _____ % CO _____ % H_2S _____ %

Calibration Performed? ☐ Yes ☐ No Initials _____

Alarm Conditions? ☐ Yes ☐ No

Monitoring Performed by (sign): _____ Date: _____ Time: _____

Continuous Air Monitoring Results

Time _____ Oxygen _____ % LEL _____ % CO _____ % H_2S _____ %

Time _____ Oxygen _____ % LEL _____ % CO _____ % H_2S _____ %

Time _____ Oxygen _____ % LEL _____ % CO _____ % H_2S _____ %

Authorization

We have reviewed the work authorized by this permit and the information contained here-in. Written instructions and safety procedures have been received and are understood. Entry cannot be approved if any checks are marked in the "NO" column. This permit is not valid unless all appropriate items are completed. This permit is to be kept at the job site. Return site copy to supervisor.

Entrant's Name _____ Signature: _____ Date: _____

Attendant's Name _____ Signature: _____ Date: _____

Supervisor's Name _____ Signature: _____ Date: _____

Confined Space Entry Permit

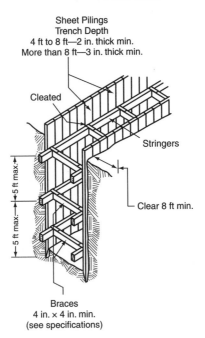

Sheet Pilings
Trench Depth
4 ft to 8 ft—2 in. thick min.
More than 8 ft—3 in. thick min.

Cleated

Stringers

5 ft max.

5 ft max.

Clear 8 ft min.

Braces
4 in. × 4 in. min.
(see specifications)

Sheet piling or equivalent solid sheeting is required for trenches 4 ft or more deep.

Longitudinal-stringer dimensions depend on the strut braces, the stringer spacing, and the depth of stringer below the ground surface.

Greater loads are encountered as the depth increases, so more or stronger stringers and struts are required near the trench bottom.

Running Material

* This section adapted from Office of Water Programs, California State University, Sacramento Foundation, in Small Water System Operation and Maintenance. For additional information, visit <www. owp.csus.edu> or call 916-278-6142.

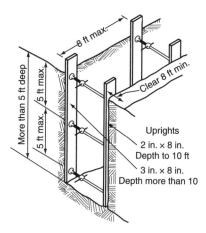

Trenches 5 ft or more deep and more than 8 ft long must be braced at intervals of 8 ft or less.

Hard Compact Ground (5 ft or more in depth)

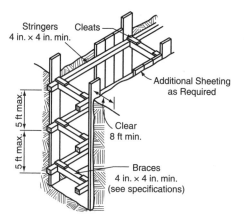

Sheeting must be provided and must be sufficient to hold the material in place.

Longitudinal-stringer dimensions depend on the strut and stringer spacing and on the degree of instability encountered.

Saturated, Filled, or Unstable Ground (additional sheeting as required)

ROADWAY, TRAFFIC, AND VEHICLE SAFETY[*]

Recommended Barricade Placement for Working in a Roadway

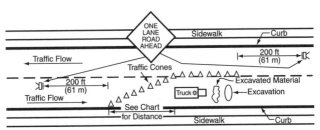

NOTE: If traffic is heavy or construction work causes interference in the open lane, one or more flaggers should be used.

Speed Limit, mph (km/hr)	Lane Width, 10 ft (3 m)		Lane Width, 11 ft (3.4 m)		Lane Width, 12 ft (3.7 m)		Minimum Number of Cones Required
	Taper Length, ft	(m)	ft	(m)	ft	(m)	
20 (32)	70	(21)	75	(23)	80	(24)	5
25 (40)	105	(32)	115	(35)	125	(38)	6
30 (48)	150	(46)	165	(50)	180	(55)	7
35 (56)	205	(62)	225	(69)	245	(75)	8
40 (64)	270	(82)	295	(90)	320	(98)	9
45 (72)	450	(137)	495	(151)	540	(165)	13
50 (81)	500	(152)	550	(168)	600	(183)	13
55 (89)	550	(168)	605	(184)	660	(201)	13

* This section adapted from Office of Water Programs, California State University, Sacramento Foundation, in Small Water System Operation and Maintenance. For additional information, visit <www. owp.csus.edu> or call 916-278-6142.

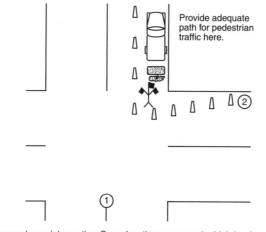

Provide adequate path for pedestrian traffic here.

Placement near intersection. Some locations may require high-level warnings at points 1 and 2.

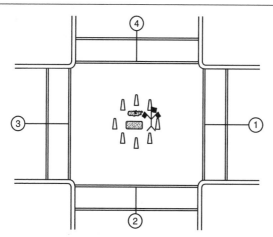

Placement at major traffic signal–controlled intersection where congestion is extreme. Some locations may permit warnings at points 1, 2, 3, and 4.

Placement of Traffic Cones and Signs

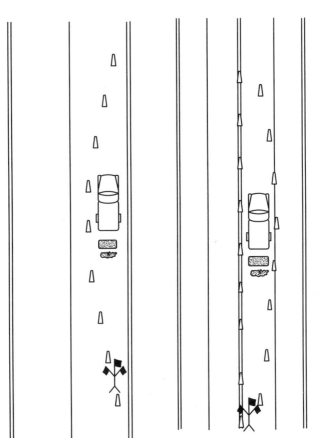

Placement for normal service, leak, or construction. See table on page 90 for distances.

Placement for multilane highway. Place high-level warning in same lane as obstruction. See table on page 90 for distances.

Placement of Traffic Cones and Signs (continued)

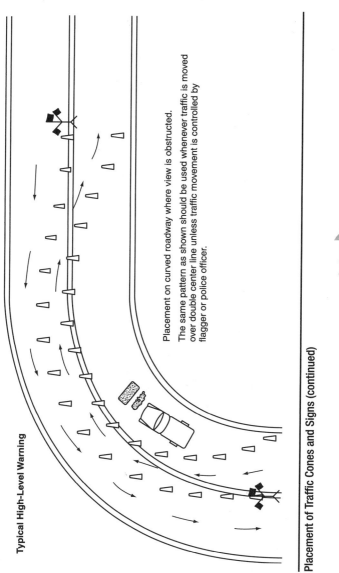

Typical High-Level Warning

Placement on curved roadway where view is obstructed.

The same pattern as shown should be used whenever traffic is moved over double center line unless traffic movement is controlled by flagger or police officer.

Placement of Traffic Cones and Signs (continued)

Safety

93

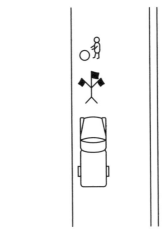

Placement for gate operation or other jobs of short duration.
Employee must wear high-visibility vest or jacket.

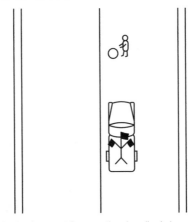

Alternate placement for operation described above.
High-level warning is mounted on rear of vehicle that is
parked in advance of work location. Employee must wear
high-visibility vest or jacket.

Placement of Traffic Cones and Signs (continued)

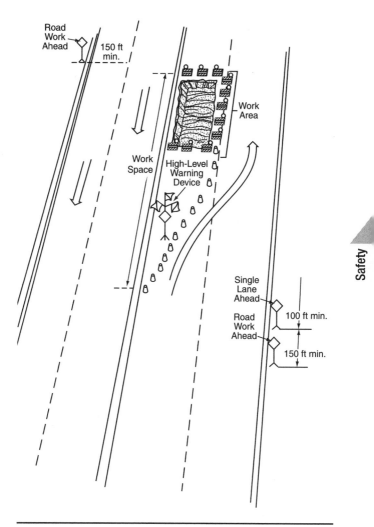

Road Work Ahead

150 ft min.

Work Area

Work Space

High-Level Warning Device

Single Lane Ahead

100 ft min.

Road Work Ahead

150 ft min.

Safety

Closing of Left Lane

95

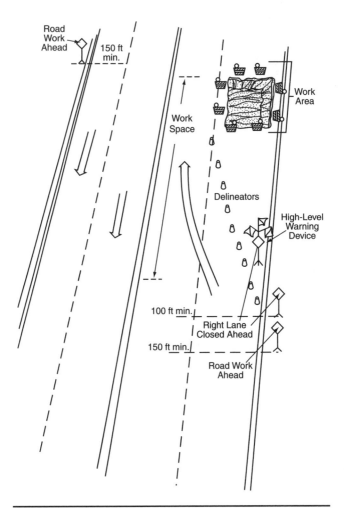

Closing of Right Lane

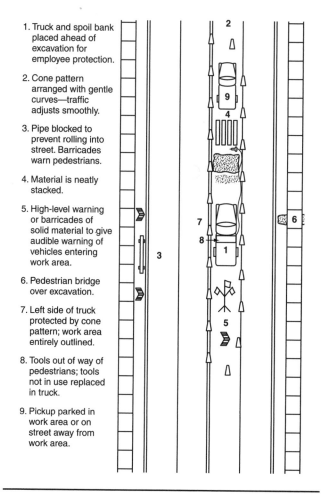

1. Truck and spoil bank placed ahead of excavation for employee protection.

2. Cone pattern arranged with gentle curves—traffic adjusts smoothly.

3. Pipe blocked to prevent rolling into street. Barricades warn pedestrians.

4. Material is neatly stacked.

5. High-level warning or barricades of solid material to give audible warning of vehicles entering work area.

6. Pedestrian bridge over excavation.

7. Left side of truck protected by cone pattern; work area entirely outlined.

8. Tools out of way of pedestrians; tools not in use replaced in truck.

9. Pickup parked in work area or on street away from work area.

Safety

Good Practices in Work Area Protection

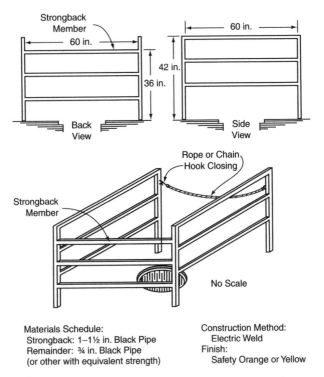

Materials Schedule:
Strongback: 1–1½ in. Black Pipe
Remainder: ¾ in. Black Pipe
(or other with equivalent strength)

Construction Method:
Electric Weld
Finish:
Safety Orange or Yellow

NOTE: Strongback member and both sides should be coupled together so they can be stacked for transportation and quickly assembled if needed.

Typical Portable Manhole Safety Enclosure

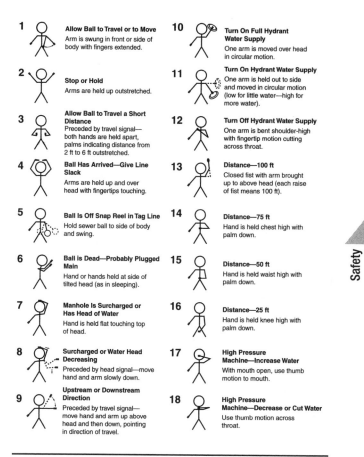

1 Allow Ball to Travel or to Move
Arm is swung in front or side of body with fingers extended.

2 Stop or Hold
Arms are held up outstretched.

3 Allow Ball to Travel a Short Distance
Preceded by travel signal—both hands are held apart, palms indicating distance from 2 ft to 6 ft outstretched.

4 Ball Has Arrived—Give Line Slack
Arms are held up and over head with fingertips touching.

5 Ball Is Off Snap Reel in Tag Line
Hold sewer ball to side of body and swing.

6 Ball is Dead—Probably Plugged Main
Hand or hands held at side of tilted head (as in sleeping).

7 Manhole Is Surcharged or Has Head of Water
Hand is held flat touching top of head.

8 Surcharged or Water Head Decreasing
Preceded by head signal—move hand and arm slowly down.

9 Upstream or Downstream Direction
Preceded by travel signal—move hand and arm up above head and then down, pointing in direction of travel.

10 Turn On Full Hydrant Water Supply
One arm is moved over head in circular motion.

11 Turn On Hydrant Water Supply
One arm is held out to side and moved in circular motion (low for little water—high for more water).

12 Turn Off Hydrant Water Supply
One arm is bent shoulder-high with fingertip motion cutting across throat.

13 Distance—100 ft
Closed fist with arm brought up to above head (each raise of fist means 100 ft).

14 Distance—75 ft
Hand is held chest high with palm down.

15 Distance—50 ft
Hand is held waist high with palm down.

16 Distance—25 ft
Hand is held knee high with palm down.

17 High Pressure Machine—Increase Water
With mouth open, use thumb motion to mouth.

18 High Pressure Machine—Decrease or Cut Water
Use thumb motion across throat.

Safety

Hand Signals in Sewer Cleaning

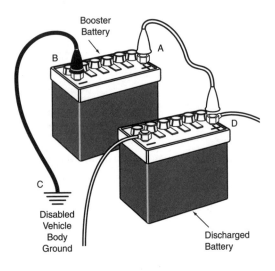

Booster
Battery

B

A

D

C

Disabled
Vehicle
Body
Ground

Discharged
Battery

Proper Booster Cable Hookup

To boost the battery of a disabled vehicle from that of another vehicle, follow this procedure.

For maximum eye safety, wear protective goggles around vehicle batteries to keep flying battery fragments and chemicals out of the eyes. Should battery acid get into the eyes, immediately flush them with water continuously for 15 minutes, then see a doctor.

First, extinguish all cigarettes and flames. A spark can ignite hydrogen gas from the battery fluid. Next, take off the battery caps, if removable, and add distilled water if it is needed. Check for ice in the battery fluid. Never jump-start a frozen battery! Replace the caps.

Next, park the vehicle with the "live" battery close enough so the cables will reach between the batteries of the two vehicles. The vehicles can be parked close, but do not allow them to touch. If they touch, this can create a dangerous situation. Now set each vehicle's parking brake. Be sure that an automatic transmission is set in park; put a manual-shift transmission in neutral. Make sure your headlights, heater, and all other electrical accessories are off (you don't want to sap electricity away from the discharged [dead] battery while you're trying to start the vehicle). If the two batteries have vent caps, remove them. Then lay a cloth over the open holes. This will reduce the risk of explosion (relieves pressure within the battery).

Attach one end of the jumper cable to the positive terminal of the booster battery (A) and the other end to the positive terminal of the discharged battery (D). The positive terminal is identified by a + sign, a red color, or a "P" on the battery in each vehicle. Each of the two booster cables has an alligator clip at each end. To attach, simply squeeze the clip, place it over the terminal, and let it shut. Now attach one end of the remaining booster cable to the negative terminal of the booster battery (B). The negative terminal is marked with a – sign, a black color, or the letter "N." Attach the other end of the cable to a metal part on the engine of the disabled vehicle (C). Many mechanics simply attach it to the negative post of the battery, but this is not recommended because a resulting arc could ignite hydrogen gas present at the battery surface and cause an explosion. Be sure that the cables do not interfere with the fan blades or belts. The engine in the booster vehicle should be running, although it is not an absolute necessity.

Get in the disabled vehicle and start the engine. After it starts, remove the booster cables. Removal is the exact reverse of attachment. Remove the black cable attached to the previously disabled vehicle, then remove it from the negative terminal of the booster battery. Next, remove the remaining cable from the positive terminal of the dead battery and then from the booster vehicle. Replace the vent caps and you're done. Have the battery and/or charging system of the vehicle checked by a mechanic to correct any problems.

FIRE AND ELECTRICAL SAFETY

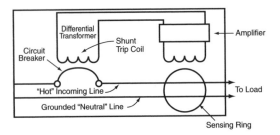

The differential transformer continuously measures the current flow in the "hot" and "neutral" lines. Under normal conditions, the current is equal in each line. If there is a difference of as little as 5 mA (0.005 A) the amplifier energizes the shunt trip coil which causes the circuit breaker to trip in 1/40th of a second or less.

EXAMPLE: A hand drill has a defective motor winding allowing a portion of the current to flow to the metal case and thus through your body causing a shock and possible electrocution.

Always use a ground fault interrupter when using electrical equipment outdoors and in damp, wet locations. Always make sure your electrical tools are in good shape.

Ground Fault Interrupter

Types of Fires and Fire Extinguishers

Combustible Material	Class of Fire and Extinguisher Marking	Extinguish With
Paper, wood, cloth	A (ordinary combustibles)	Water, soda-acid, and dry chemical rated A, B, C
Oil, tar, gasoline, paint	B (flammable liquids)	Foam, carbon dioxide, liquid gas (Halon™), and dry chemical rated B, C, or A, B, C
Electric motors, power cords, wiring, and transformer boxes	C (electrical equipment)	Carbon dioxide, liquid gas (Halon™), and dry chemical rated B, C, or A, B, C
Sodium, zinc phosphorus, magnesium, potassium, and titanium, especially as dust or turnings	D (special metals)	Only special dry-powder extinguishers marked for this purpose

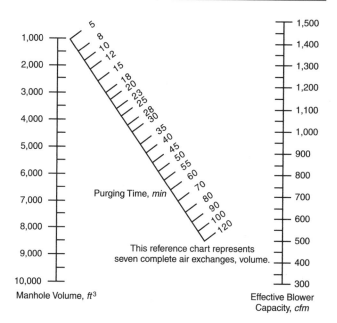

Manhole Volume, ft^3

Effective Blower Capacity, cfm

This reference chart represents seven complete air exchanges, volume.

Purging Time, *min*

Safety

Use of alignment chart:
1. Place straightedge on manhole volume (left scale).
2. Place either end of straightedge on blower capacity (right scale).
3. Read required purging time, in minutes, on diagonal scale.
4. If two blowers are used, add the two capacities, then proceed as above.
5. When common gases are encountered, increase purging time by 50%.
6. Effective blower capacity is measured with one or two 90° bends in standard 15-ft blower hose.

Ventilation Nomograph

Hazardous Location Information

A hazardous location is an area where the possibility of explosion and fire is created by the presence of flammable gases, vapors, dusts, fibers, or flyings. (Fibers and flyings are not likely to be suspended in the air but can collect around machinery or on lighting fixtures where heat, a spark, or hot metal can ignite them.)

Class I

(National Electrical Code [NEC]-500-5)

Those areas in which flammable gases or vapors may be present in the air in sufficient quantities to be explosive or ignitable.

Class II

(NEC-500-6)

Those areas made hazardous by the presence of combustible dust.

Class III

(NEC-500-7)

Those areas in which there are easily ignitable fibers or flyings present due to type of material being handled, stored, or processed.

Division I

(NEC-800-5, 6, 7)

In the normal situation, hazard would be expected to be present in everyday production operations or during frequent repair and maintenance activity.

Division II

(NEC-500-5, 6, 7)

In the abnormal situation, material is expected to be confined within closed containers or closed systems and will be present only through accidental rupture, breakage, or unusual faulty operation.

Groups

(NEC-500-3)

The gases and vapors of class locations are broken into four groups by the code: A, B, C, and D. These materials are grouped according to the ignition temperature of the substance, its explosion pressure, and other flammable characteristics. The dust locations of Class II are designated E, F, and G. These groups are classified according to the ignition temperature and the conduction of the hazardous substance.

NOTE: For detailed group descriptions, refer to NEC-500-3.

Table continued on next page

Hazardous Location Information (continued)

Typical Class I Locations

- Petroleum refineries, and gasoline storage and dispensing areas
- Industrial firms that use flammable liquids in dip tanks for parts cleaning or other operations
- Petrochemical companies that manufacture chemicals from gas and oil
- Dry-cleaning plants where vapors from cleaning fluids can be present
- Companies that have spraying areas where products are coated with paint or plastics
- Aircraft hangars and fuel servicing areas
- Utility gas plants, and operations involving storage and handling of liquefied petroleum gas or natural gas

Typical Class II Locations

- Grain elevators, flour and feed mills
- Plants that manufacture, use, or store magnesium or aluminum powders
- Plants that have chemical or metallurgical processes; producers of plastics, medicines, and fireworks
- Producers of starch or candies
- Spice-grinding plants, sugar plants, and cocoa plants
- Coal preparation plants and other carbon-handling or processing areas

Typical Class III Locations

- Textile mills, cotton gins, cotton seed mills, and flax processing plants
- Any plant that shapes, pulverizes, or cuts wood and creates sawdust or flyings

Source: Explosion Proof Blowers: 9503 and 9515-01 NEC. (Warning: Explosion-proof blowers must be used with statically conductive ducting.)

Safety

Hazards Classification

Class 1	Explosives
Class 2	Gas
Class 3	Flammable liquid
Class 4	Flammable solids (potential spontaneous combustion, or emission of flammable gases when in contact with water)
Class 5	Oxidizing substances and organic peroxides
Class 6	Toxic (poisonous) and infectious substances
Class 7	Radioactive material
Class 8	Corrosives
Class 9	Miscellaneous dangerous goods

HEALTH EFFECTS OF TOXIN EXPOSURE

Although the foul odor (rotten eggs) of hydrogen sulfide is easily detected at low concentrations, it is an unreliable warning because the gas rapidly desensitizes the sense of smell, leading to a false sense of security. In high concentrations of hydrogen sulfide, a worker may collapse with little or no warning.

Potential Effects of Hydrogen Sulfide Exposure

ppm	Effects and Symptoms	Time
1,000 or more	Unconsciousness, death	Minutes
500–700	Unconsciousness, death	30 minutes to 1 hour
200–300	Marked eye and respiratory irritations	1 hour
50–100	Mild eye and respiratory irritations	1 hour
10	Permissible exposure level	8 hours

Carbon monoxide is an odorless, colorless gas that may build up in a confined space. In high concentrations of carbon monoxide a worker may collapse with little or no warning.

Potential Effects of Carbon Monoxide Exposure

ppm	Effects and Symptoms	Time
4,000	Fatal	<1 hour
2,000–2,500	Unconsciousness	30 minutes
1,000–2,000	Slight heart palpitation	30 minutes
1,000–2,000	Tendency to stagger	1½ hours
1,000–2,000	Confusion, headache, nausea	2 hours
600	Headache, discomfort	1 hour
400	Headache, discomfort	2 hours
200	Slight headache, discomfort	3 hours
50	Permissible exposure limit	8 hours

Chlorine is a highly toxic chemical even in small concentrations in air. The following table shows the physiological effects of various concentrations of chlorine by volume in air.

Effects of Chlorine Gas Exposure

ppm	Effects and Symptoms
1	Slight symptoms after several hours' exposure
3	Detectable odor
4	60-minute inhalation without serious effects
5	Noxiousness
15	Throat irritation
30	Coughing
40	Dangerous from 30 minutes to 1 hour
1,000	Death after a few deep breaths

Safety

Common Dangerous Gases Encountered in Water Supply Systems and

Name of Gas	Chemical Formulae	Specific Gravity of Vapor Density* (air = 1)	Explosive Range (% by volume in air)	
			Lower Limit	Upper Limit
Carbon dioxide	CO_2	1.53	Not flammable	Not flammable
Carbon monoxide	CO	0.97	12.5	74.2
Chlorine	Cl_2	2.5	Not flammable Not explosive	Not flammable Not explosive
Ethane	C_2H_4	1.05	3.1	15.0
Gasoline vapor	C_5H_{12} to C_9H_{20}	3.0 to 4.0	1.3	7.0
Hydrogen	H_2	0.07	4.0	74.2

Common Properties (percentages given are percent in air by volume)	Physiological Effects (percentages given are percent in air by volume)	Most Common Sources in Sewers	Simplest and Least Expensive Safe Method of Testing[†]
Colorless, odorless, nonflammable. Not generally present in dangerous amounts unless there is already an oxygen deficiency.	10% cannot be endured for more than a few minutes. Acts on nerves of respiration.	Issues from carbonaceous strata. Sewer gas.	Oxygen deficiency indicator
Colorless, odorless, nonirritating, tasteless. Flammable. Explosive.	Hemoglobin of blood has strong affinity for gas causing oxygen starvation. 0.2% to 0.25% causes unconsciousness in 30 minutes.	Manufactured fuel gas	CO ampoules
Greenish yellow gas, or amber color liquid under pressure. Highly irritating and penetrating odor. Highly corrosive in presence of moisture.	Respiratory irritant, irritating to eyes and mucous membranes. 30 ppm causes coughing. 40–60 ppm dangerous in 30 minutes. 1,000 ppm likely to be fatal in a few breaths.	Leaking pipe connections. Overdosage.	Chlorine detector. Odor, strong. Ammonia on swab gives off white fumes.
Colorless, tasteless, odorless, nonpoisonous. Flammable. Explosive.	See Hydrogen.	Natural gas	Combustible gas indicator
Colorless. Odor noticeable in 0.03%. Flammable. Explosive.	Anesthetic effects when inhaled. 2.43% rapidly fatal. 1.1% to 2.2% dangerous for even short exposure.	Leaking storage tanks, discharges from garages, and commercial or home dry-cleaning operations.	1. Combustible gas indicator 2. Oxygen deficiency indicator for concentrations >30%
Colorless, odorless, tasteless, nonpoisonous. Flammable. Explosive. Propagates flame rapidly; very dangerous.	Acts mechanically to deprive tissues of oxygen. Does not support life. A simple asphyxiant.	Manufactured fuel gas	Combustible gas indicator

Safety

Table continued on next page

109

Common Dangerous Gases Encountered in Water Supply Systems and

Name of Gas	Chemical Formulae	Specific Gravity of Vapor Density* (air = 1)	Explosive Range (% by volume in air)	
			Lower Limit	Upper Limit
Hydrogen sulfide	H_2S	1.19	4.3	46.0
Methane	CH_4	0.55	5.0	15.0
Nitrogen	N_2	0.97	Not flammable	Not flammable
Oxygen (in air)	O_2	1.11	Not flammable	Not flammable

* Gases with a specific gravity less than 1.0 are lighter than air; those with a specific gravity more than 1.0 are heavier than air.
† The first method given is the preferable testing procedure.
‡ Never enter a 12% atmosphere. Use detection meters with alarm warning devices.

Common Properties (percentages below are percent in air by volume)	Physiological Effects (percentages below are percent in air by volume)	Most Common Sources in Sewers	Simplest and Cheapest Safe Method of Testing[†]
Rotten egg odor in small concentrations but sense of smell rapidly impaired. Odor not evident at high concentrations. Colorless. Flammable. Explosive. Poisonous.	Death in a few minutes at 0.2%. Paralyzes respiratory center.	Petroleum fumes, from blasting, sewer gas	1. H_2S analyzer 2. H_2S ampoules
Colorless, tasteless, odorless, nonpoisonous. Flammable. Explosive.	See Hydrogen.	Natural gas, marsh gas, manufacturing fuel gas, sewer gas	1. Combustible gas indicator 2. Oxygen deficiency indicator
Colorless, tasteless, odorless. Nonflammable. Nonpoisonous. Principal constituent of air (about 79%).	See Hydrogen.	Issues from some rock strata. Sewer gas	Oxygen deficiency indicator
Colorless, odorless, tasteless, nonpoisonous gas. Supports combustion.	Normal air contains 20.93% of O_2. Humans tolerate down to 12%.[‡] Below 5% to 7%, likely to be fatal.	Oxygen depletion from poor ventilation and absorption or chemical consumption of available O_2.	Oxygen deficiency indicator

Safety

Chlorine and Safety

When using chlorine, observe the following precautions:

1. Use a mask when entering a chlorine-containing atmosphere.

2. Apparatus, lines, and cylinder valves should be checked regularly for leaks. Use ammonia fumes to test leaks. Ammonia and chlorine produce white fumes of ammonium chloride, which indicate leaks.

3. Because it is heavier than air, always store chlorine on the lowest floor; it will collect at the lower level. For the same reason, never stoop down when a chlorine smell is noticed.

Handle chlorine carefully and respectfully, as she is the "green goddess of water."

Waterborne Diseases

Waterborne Disease	Causative Organism	Source of Organism in Water	Symptom/Outcome
Gastroenteritis	*Salmonella* (bacteria)	Animal or human feces	Acute diarrhea and vomiting—rarely fatal
Typhoid	*Salmonella typhosa* (bacteria)	Human feces	Inflamed intestine, enlarged spleen, high temperature—fatal
Dysentery	*Shigella*	Human feces	Diarrhea—rarely fatal
Cholera	*Vibrio cholerae* (bacteria)	Human feces	Vomiting, severe diarrhea, rapid dehydration, mineral loss—high mortality
Infectious hepatitis	Virus	Human feces, shellfish grown in polluted waters	Yellowed skin, enlarged liver, abdominal pain; lasts as long as 4 months—low mortality
Amoebic dysentery	*Entamoeba histolytica* (protozoa)	Human feces	Mild diarrhea, chronic dysentery—rarely fatal
Giardiasis	*Giardia lamblia* (protozoa)	Wild animal feces suspected	Diarrhea, cramps, nausea, general weakness; lasts 1 to 30 weeks—not fatal
Cryptosporidiosis	*Cryptosporidium*	Human and animal feces	Diarrhea, abdominal pain, vomiting, low-grade fever—rarely fatal

Safety

Potential Waterborne Disease-Causing Organisms

Organism	Major Disease	Primary Source
Bacteria		
Salmonella typhi	Typhoid fever	Human feces
Salmonella paratyphi	Paratyphoid fever	Human feces
Other *Salmonella* spp.	Gastroenteritis (salmonellosis)	Human/animal feces
Shigella	Bacillary dysentery	Human feces
Vibrio cholerae	Cholera	Human feces, coastal water
Pathogenic *Escherichia coli*	Gastroenteritis	Human/animal feces
Yersinia enterocolitica	Gastroenteritis	Human/animal feces
Campylobacter jejuni	Gastroenteritis	Human/animal feces
Legionella pneumophila	Legionnaires' disease, Pontiac fever	Warm water
Mycobacterium avium intracellulare	Pulmonary disease	Human/animal feces, soil, water
Pseudomonas aeruginosa	Dermatitis	Natural waters
Aeromonas hydrophila	Gastroenteritis	Natural waters
Helicobacter pylori	Peptic ulcers	Saliva, human feces suspected
Enteric Viruses		
Poliovirus	Poliomyelitis	Human feces
Coxsackievirus	Upper respiratory disease	Human feces

Table continued on next page

114

Potential Waterborne Disease-Causing Organisms (continued)

Organism	Major Disease	Primary Source
Echovirus	Upper respiratory disease	Human feces
Rotavirus	Gastroenteritis	Human feces
Norwalk virus and other calciviruses	Gastroenteritis	Human feces
Hepatitis A virus	Infectious hepatitis	Human feces
Hepatitis E virus	Hepatitis	Human feces
Astrovirus	Gastroenteritis	Human feces
Enteric adenoviruses	Gastroenteritis	Human feces
Protozoa and Other Organisms		
Giardia lamblia	Giardiasis (gastroenteritis)	Human and animal feces
Cryptosporidium parvum	Cryptosporidiosis (gastroenteritis)	Human and animal feces
Entamoeba histolytica	Amoebic dysentery	Human feces
Cyclospora cayatanensis	Gastroenteritis	Human feces
Microspora	Gastroenteritis	Human feces
Acanthamoeba	Eye infection	Soil and water
Toxoplasma gondii	Flu-like symptoms	Cats
Naegleria fowleri	Primary amoebic meningoencephalitis	Soil and water
Blue-green algae	Gastroenteritis, liver damage, nervous system damage	Natural waters
Fungi	Respiratory allergies	Air, water suspected

Typical Pathogen Survival Times at 20°–30°C

| | Survival Time, *days* | | |
Pathogen	Fresh Water and Sewage	Crops	Soil
Viruses[*]			
Enteroviruses[†]	<120 but usually <50	<60 but usually <15	<100 but usually <20
Bacteria			
Fecal coliforms[*]	<60 but usually <30	<30 but usually <15	<70 but usually <20
Salmonella spp.[*]	<60 but usually <30	<30 but usually <15	<70 but usually <20
Shigella spp.[*]	<30 but usually <10	<10 but usually <5	
Vibrio cholerae[‡]	<30 but usually <10	<5 but usually <2	<20 but usually <10
Protozoa			
E. histolytica cysts	<30 but usually <15	<10 but usually <2	<20 but usually <10
Helminths			
A. lumbricoides eggs	Many months	<60 but usually <30	Many months

[*] In seawater, viral survival is less, and bacterial survival is very much less, than in freshwater.
[†] Includes polio-, echo-, and coxsackie viruses.
[‡] *V. cholerae* survival in aqueous environments is a subject of current uncertainty.

Infectious Doses of Selected Pathogens

Organisms	Infectious Dose[*]
Escherichia coli (enteropathogenic)	10^6–10^{10}
Clostridium perfringens	1–10^{10}
Salmonella typhi	10^4–10^7
Vibrio cholerae	10^3–10^7
Shigella flexneri 2A	180
Entamoeba histolytica	20
Shigella dysentariae	10
Giardia lamblia	<10
Cryptosporidium parvum	1–10
Ascaris lumbricoides	1–10
Enteric virus	1–10

[*] Some of the data for bacteria are given as ID_{50}, which is the dose that infects 50% of the people given that dose. People given lower doses also could become infected.

Safety

Microorganism Concentrations in Raw Wastewater

Organisms	Concentration, *number/100 mL*
Total coliforms	10^7–10^{10}
Clostridium perfringens	10^3–10^5
Enterococci	10^4–10^5
Fecal coliforms	10^4–10^9
Fecal *Streptococci*	10^4–10^6
Pseudomonas aeruginosa	10^3–10^4
Protozoan cysts	10^3–10^5
Shigella	1–10^3
Salmonella	10^2–10^4
Helminth ova	10–10^3
Enteric virus	10^2–10^4
Giardia lamblia cysts	10–10^4
Entamoeba histolytica cysts	1–10
Cryptosporidium parvum oocysts	10–10^3

Examples of Concentration of Microbial Pathogens in Raw Wastewater and Sludge

Microbial Agent	Raw Wastewater, number/L	Sludge, number/gm
Salmonella	4×10^3 MPN	2×10^3 MPN
Enteric virus	3×10^4 pfu	1×10^3 pfu
Giardia	2×10^2 cysts	1×10^2 cysts
Cryptosporidium	2×10^2 oocysts	ND[*]
Helminths	8×10^2 ova	3×10 ova

[*] ND = no data.

Estimate of Percent Removal of Selected Microbial Pathogens Using Conventional Treatment Processes

Microbial Agent	Primary Treatment	Secondary Treatment	Digested Sludge
Salmonella	50	99	99
Enteric virus	70	99	15
Giardia cysts	50	75	30
Helminth ova	90	99.99	30

Collection

Miles of pipes connect homes to wastewater treatment plants. Some are gravity systems and some are pressure systems. These systems must operate properly to protect public health and the environment.

DESIGN FLOW RATES

The average daily flow (volume per unit time), maximum daily flow, peak hourly flow, minimum hourly and daily flows, and design peak flow are generally used as the basis of design for sewers, lift stations, sewage (wastewater) treatment plants, treatment units, and other wastewater handling facilities. Definitions and purposes of flow are given as follows.

The design average flow is the average of the daily volumes to be received for a continuous 12-month period of the design year. The average flow may be used to estimate pumping and chemical costs, sludge generation, and organic-loading rates.

The maximum daily flow is the largest volume of flow to be received during a continuous 24-hour period. It is employed in the calculation of retention time for equalization basin and chlorine contact time.

The peak hourly flow is the largest volume received during a 1-hour period, based on annual data. It is used for the design of collection and interceptor sewers, wet wells, wastewater pumping stations, wastewater flow measurements, grit chambers, settling basins, chlorine contact tanks, and pipings. The design peak flow is the instantaneous maximum flow rate to be received. The peak hourly flow is commonly assumed to be three times the average daily flow.

The minimum daily flow is the smallest volume of flow received during a 24-hour period. The minimum daily flow is important in the sizing of conduits where solids might be deposited at low-flow rates.

The minimum hourly flow is the smallest hourly flow rate occurring over a 24-hour period, based on annual data. It is important to the sizing of wastewater flowmeters, chemical-feed systems, and pumping systems.

Example

Estimate the average and maximum hourly flow for a community of 10,000 persons.

Step 1. Estimate wastewater daily flow rate.

Assume average water consumption = 200 L/(capita·day)

Assume 80% of water consumption goes to the sewer.

$$\text{average wastewater flow} = 200 \text{ L/(c·d)} \times 0.80$$
$$\times 10,000 \text{ persons} \times 0.001 \text{ m}^3/\text{L}$$
$$= 1,600 \text{ m}^3/\text{day}$$

Step 2. Compute average hourly flow rate.

$$\text{average hourly flow rate} = 1,600 \text{ m}^3/\text{day} \times 1 \text{ day}/24 \text{ hr}$$
$$= 66.67 \text{ m}^3/\text{hr}$$

Step 3. Estimate the maximum hourly flow rate.

Assume the maximum hourly flow rate is three times the average hourly flow rate, thus

$$\text{maximum hourly flow rate} = 66.67 \text{ m}^3/\text{hr} \times 3$$
$$= 200 \text{ m}^3/\text{hr}$$

Collection

Minimum Slopes for Various Sized Sewers at a Flowing Full Velocity of 2.0 ft/sec and Corresponding Discharges*

Sewer Diameter, in.	Minimum Slope, ft/100 ft	Flowing Full Discharge	
		ft³/sec	gpm
8	0.33	0.7	310
10	0.25	1.1	490
12	0.19	1.6	700
15	0.14	2.4	1,080
18	0.11	3.5	1,570
21	0.092	4.8	2,160
24	0.077	6.3	2,820
27	0.066	8.0	3,570
30	0.057	9.8	4,410
36	0.045	14.1	6,330

Courtesy of Pearson Education, Inc.

* Based on Manning's formula with $n = 0.013$.

Velocity Formula

$$\text{velocity, ft/sec} = \frac{\text{distance traveled, ft}}{\text{time of test, sec}}$$

Area of Partly Filled Circular Pipes

d/D	Factor	d/D	Factor	d/D	Factor	d/D	Factor
0.01	0.0013	0.26	0.1623	0.51	0.4027	0.76	0.6405
0.02	0.0037	0.27	0.1711	0.52	0.4127	0.77	0.6489
0.03	0.0069	0.28	0.1800	0.53	0.4227	0.78	0.6573
0.04	0.0105	0.29	0.1890	0.54	0.4327	0.79	0.6655
0.05	0.0174	0.30	0.1982	0.55	0.4426	0.80	0.6736
0.06	0.0192	0.31	0.2074	0.56	0.4526	0.81	0.6815
0.07	0.0242	0.32	0.2167	0.57	0.4625	0.82	0.6893
0.08	0.0294	0.33	0.2260	0.58	0.4724	0.83	0.6969
0.09	0.0350	0.34	0.2355	0.59	0.4822	0.84	0.7043
0.10	0.0409	0.35	0.2450	0.60	0.4920	0.85	0.7115
0.11	0.0470	0.36	0.2545	0.61	0.5018	0.86	0.7186
0.12	0.0534	0.37	0.2642	0.62	0.5115	0.87	0.7254
0.13	0.0600	0.38	0.2739	0.63	0.5212	0.88	0.7320
0.14	0.0668	0.39	0.2836	0.64	0.5308	0.89	0.7384
0.15	0.0739	0.40	0.2934	0.65	0.5404	0.90	0.7445
0.16	0.0811	0.41	0.3032	0.66	0.5499	0.91	0.7504
0.17	0.0885	0.42	0.3130	0.67	0.5594	0.92	0.7560
0.18	0.0961	0.43	0.3229	0.68	0.5687	0.93	0.7612
0.19	0.1039	0.44	0.3328	0.69	0.5780	0.94	0.7662
0.20	0.1118	0.45	0.3428	0.70	0.5872	0.95	0.7707
0.21	0.1199	0.46	0.3527	0.71	0.5964	0.96	0.7749
0.22	0.1281	0.47	0.3627	0.72	0.6054	0.97	0.7785
0.23	0.1365	0.48	0.3727	0.73	0.6143	0.98	0.7816
0.24	0.1449	0.49	0.3827	0.74	0.6231	0.99	0.7841
0.25	0.1535	0.50	0.3927	0.75	0.6319	1.00	0.7854

d = depth, inches
D = diameter, inches

FLOW MEASUREMENT

Collection system operators need to know the fundamentals of wastewater flow measurement in a sewer pipe. There are many devices available for flow measurement. All of these flow meters are based on the simple principle that the flow rate equals the velocity of flow multiplied by the cross-sectional areas of the flow. This principle is expressed by the following formula:

$$Q, \text{cubic feet per second} = (\text{area, ft}^2)(\text{velocity, ft}^2)$$

Calculation of the cross-sectional area of flow in a sewer line can be made by using a factor found in the table on page 122. This procedure is explained in the example below.

Example

The depth of flow in a 12-in. diameter sewer is 5 in. Determine the cross-sectional area of the flow.

Known	Unknown
D or diameter, in. = 12 in.	Cross-sectional area, ft^2
d or depth, in. = 5 in.	

To determine the cross-sectional area for a sewer pipe flowing partially full, use the following steps:

1. Find the value for the depth, d, divided by the diameter, D.

$$\frac{d, \text{in.}}{D, \text{in.}} = \frac{5 \text{ in.}}{12 \text{ in.}}$$

$$= 0.42 \text{ in.}$$

2. Find the correct factor for 0.42 in the table on page 122.

$$\frac{d}{D} = 0.42 \qquad \text{factor} = 0.3130 \text{ (number unknown)}$$

3. Calculate the cross-sectional area.

$$\text{Pipe cross-sectional area, sq ft} = \frac{(\text{factor})(\text{diameter, in.})^2}{144 \text{ in.}^2/\text{ft}^2}$$

$$= \frac{(0.3130)(12 \text{ in.})^2}{144 \text{ in.}^2/\text{ft}^2}$$

$$= 0.313 \text{ ft}^2$$

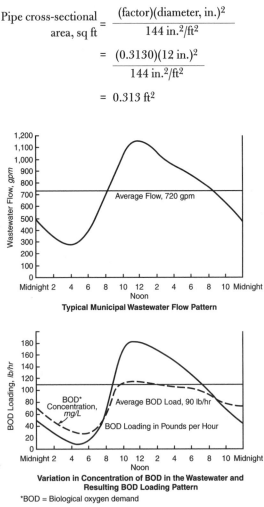

Typical Municipal Wastewater Flow Pattern

Variation in Concentration of BOD in the Wastewater and Resulting BOD Loading Pattern

*BOD = Biological oxygen demand

Courtesy of Pearson Education, Inc.

Wastewater Flow and Strength Variations for a Typical Medium-Sized City

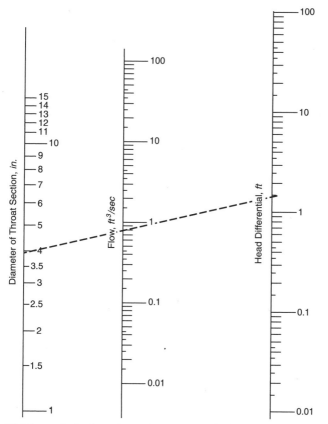

NOTE: In an actual water system environment, correction factors may be needed in the use of this nomograph.

Flow Rate Nomograph for Venturi Meter

Collection

Typical Wastewater Flow Rates for Miscellaneous Facilities

Type of Establishment	Gallons per Person per Day (unless otherwise noted)
Airports (per passenger)	5
Bathhouses and swimming pools	10
Camps	
Campground with central comfort station	35
With flush toilets, no showers	25
Construction camps (semipermanent)	50
Day camps (no meals served)	15
Resort camps (night and day) with limited plumbing	50
Luxury camps	100
Cottages and small dwellings with seasonal occupancy	75
Country clubs (per resident member)	100
Country clubs (per nonresident member present)	25
Dwellings	
Boarding houses	50
(additional for nonresident boarders)	10
Rooming houses	40
Factories (gallons per person, per shift, exclusive of industrial wastes)	35
Hospitals (per bed space)	250
Hotels with laundry (two persons per room) per room	150
Institutions other than hospitals including nursing homes (per bed space)	125
Laundries—self-service (gallons per wash)	30
Motels (per bed) with laundry	50
Picnic parks (toilet wastes only per park user)	5
Picnic parks with bathhouses, showers and flush toilets (per park user)	10
Restaurants (toilet and kitchen wastes per patron)	10
Restaurants (kitchen wastes per meal served)	3
Restaurants (additional for bars and cocktail lounges)	2
Schools	
Boarding	100
Day (without gyms, cafeterias, or showers)	15
Day (with gyms, cafeterias, and showers)	25
Day (with cafeterias, but without gyms or showers)	20
Service stations (per vehicle served)	5
Theaters	
Movie (per auditorium seat)	5
Drive-in (per car space)	10

Table continued on next page

Typical Wastewater Flow Rates for Miscellaneous Facilities (continued)

Type of Establishment	Gallons per Person per Day (unless otherwise noted)
Travel trailer parks without individual water and sewer hookups (per space)	50
Travel trailer parks with individual water and sewer hookups (per space)	100
Workers	
Offices, schools, and business establishments (per shift)	15

Approximate Wastewater Flows for Various Kinds of Establishments and Services

Type	Gallons per Person per Day	Pounds of Biological Oxygen Demand per Person per Day
Domestic wastewater from residential areas		
Large single-family houses	120	0.20
Typical single-family houses	80	0.17
Multiple-family dwellings (apartments)	60–75	0.17
Small dwellings or cottages	50	0.17
Domestic wastewater from camps and motels		
Luxury resorts	100–150	0.20
Mobile home parks	50	0.17
Tourist camps or trailer parks	35	0.15
Hotels and motels	50	0.10
Schools		
Boarding schools	75	0.17
Day schools with cafeterias	20	0.06
Day schools without cafeterias	15	0.04
Restaurants		
Each employee	30	0.10
Each patron	7–10	0.04
Each meal served	4	0.03
Transportation terminals		
Each employee	15	0.05
Each passenger	5	0.02
Hospitals	150–300	0.30
Offices	15	0.05
Drive-in theaters (per stall)	5	0.02
Movie theaters (per seat)	3–5	0.02
Factories, exclusive of industrial and cafeteria wastes	15–30	0.05

Courtesy of Pearson Education, Inc.

Collection

Average Characteristics of Selected Industrial Wastewaters

	Milk Processing	Meat Packing	Synthetic Textile	Chlorophenolic Manufacture
Biological oxygen demand, mg/L	1,000	1,400	1,500	4,300
Chemical oxygen demand, mg/L	1,900	2,100	3,300	5,400
Total solids, mg/L	1,600	3,300	8,000	53,000
Suspended solids, mg/L	300	1,000	2,000	1,200
Nitrogen, mg N/L	50	150	30	0
Phosphorus, mg P/L	12	16	0	0
pH	7	7	5	7
Temperature, °C	29	28	—	17
Grease, mg/L	—	500	—	—
Chloride, mg/L	—	—	—	27,000
Phenols, mg/L	—	—	—	140

Courtesy of Pearson Education, Inc.

SEWER CONSTRUCTION

Conduit material for sewer construction consists of two types: rigid pipe and flexible pipe. Specified rigid materials include asbestos–cement, cast iron, concrete, and vitrified clay. Flexible materials include ductile iron, fabricated steel, corrugated aluminum, thermoset plastic (reinforced plastic mortar and reinforced thermosetting resin), and thermoplastic. Thermoplastic consists of acrylonitrile–butadiene–styrene (ABS), ABS composite, polyethylene (PE), and polyvinyl chloride (PVC).

Nonpressure sewer pipe is commercially available in the size range from 4 to 42 in. (102 to 1,067 mm) in diameter and 13 ft (4.0 m) in length. Half-length sections of 6.5 ft (2 m) are available for smaller size pipes.

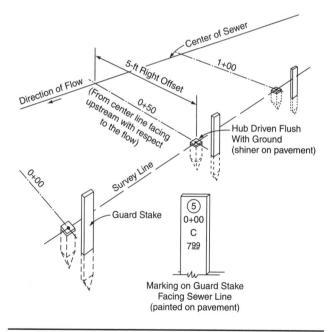

Collection

Control Points for Sewer Construction (continued on next page)

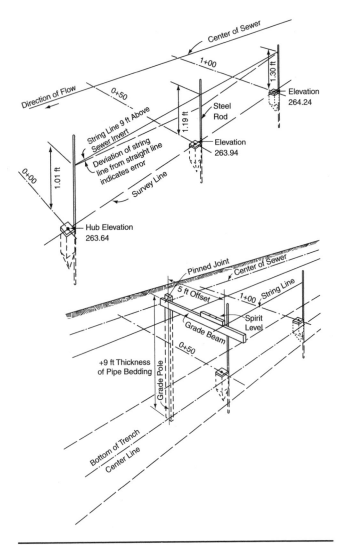

Control Points for Sewer Construction (continued)

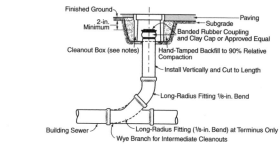

Terminate cleanout at closest joint to surface with temporary plug. After all backfill is complete and subgrade made in areas to be paved, the final riser pipe and box shall be installed as shown.

Cleanout at Property Line

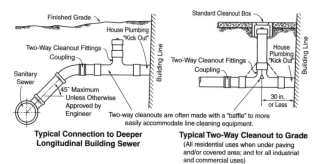

Typical Connection to Deeper Longitudinal Building Sewer

Two-way cleanouts are often made with a "baffle" to more easily accommodate line cleaning equipment.

Typical Two-Way Cleanout to Grade
(All residential uses when under paving and/or covered area; and for all industrial and commercial uses)

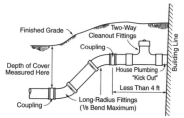

Typical Connection to Building Sewer Where Additional Depth Is Required

NOTES: 1. Cleanouts should be extended to surface so they are accessible without excavation in order to reduce maintenance costs and customer complaints regarding yard disturbance.

2. It may be difficult to push equipment through two-way cleanout fittings because of the right-angle entrance instead of a gradual entrance.

Types and Locations of Building Sewer Cleanouts

Collection

Low-Pressure Collection System

Where the topography and ground conditions of an area are not suitable for a conventional gravity collection system due to flat terrain, rocky conditions, or extremely high groundwater, low-pressure collection systems are now becoming a practical alternative. Pressure sewers may be installed instead of gravity sewers in an area because (1) a pipe slope is not practical to maintain gravity flow, (2) smaller pipe sizes can be used due to pressurization, and (3) reduced pipe sizes can be installed due to a lack of infiltration and inflow because the pipeline has no leaks and water does not enter the system through manholes. Operation and maintenance considerations when comparing pressure sewers with gravity systems include the facts that pressure systems have (1) higher energy costs for pumping; (2) greater costs for pumping facilities; (3) fewer stoppages; (4) no root intrusion; (5) no extra capacity for infiltration and inflow; (6) no deep trenches or buried pipe; and (7) no inverted siphons for crossing roads or rivers. The principal components of a low-pressure collection system include gravity sewers, holding tanks, grinder pumps, and pressure mains.

Gravity sewers connect a building's wastewater drainage system to a buried pressurization unit (containing a holding tank) located on the lot as illustrated in the accompanying figure.

Holding tanks serve as a reservoir for grinder pumps and have a capacity of approximately 50 gallons. The figure also illustrates a typical pressurization unit with a holding tank.

Grinder pumps serve both as a unit to grind the solids in the wastewater (that could plug the downstream small-diameter pressure sewers and valves) and to pressurize the wastewater to help move it through the collection system. The figure also illustrates the location of the submersible grinder pump in the holding tank.

Pressure mains are the "arteries" of the low-pressure collection system and convey the pressurized wastewater to a treatment plant. Because the wastewater is "pushed" by pressure, the mains

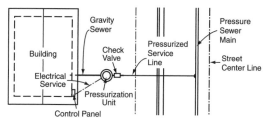

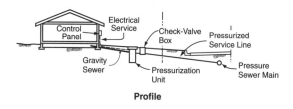

Pressurization Unit Contains: Holding Tank, Grinder Pump, Float Switches, and Gate Valve

Plan

Profile

Principal Components of a Typical Low-Pressure Collection System

are not dependent on a slope to create a gravity flow and can be laid at a uniform depth following the natural slope of the land along their routes. Low-pressure collection systems must have access for maintenance. This means line access where a pig can be inserted into a line for cleaning and also removed from the line. "Pig" refers to a poly pig, which is a bullet-shaped device made of hard rubber or similar material. Manholes or boxes must have valves and pipe spools (2- to 3-ft-long flanged sections of pipe) that can be removed for cleaning the pipe or for pumping into or out of the system with a portable pump. Refer to the figures that illustrate the profile of a typical low-pressure main and a typical low-pressure collection system.

Collection

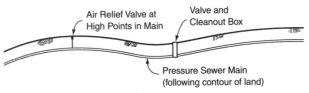

NOTE: Vertical scale is exaggerated.

Profile of a Typical Low-Pressure Main

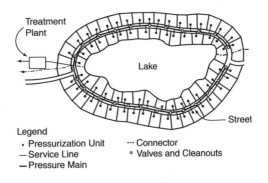

Schematic of a Typical Low-Pressure Collection System

Vacuum Collection Systems

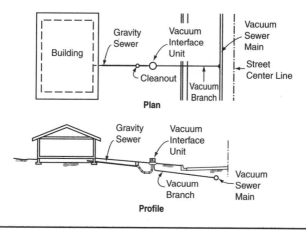

Principal Components of a Typical Vacuum Collection System

NOTE: Vertical scale is exaggerated.

Profile of a Typical Vacuum Collection System

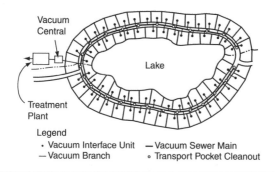

Legend
- Vacuum Interface Unit
- Vacuum Branch
- Vacuum Sewer Main
- Transport Pocket Cleanout

Schematic of a Typical Vacuum Collection System

Collection

Backfill Loads in Pounds per Linear Foot on 8-in. Circular Pipe in a Trench Installation Based on 100 lb/ft³ Ordinary Clay Fill

Height of Backfill H Above Top of Pipe, ft	Trench Width at Top of Pipe										Transition Width†
	1 ft 6 in.	1 ft 9 in.	2 ft 0 in.	2 ft 3 in.	2 ft 6 in.	2 ft 9 in.	3 ft 0 in.	3 ft 3 in.	3 ft 6 in.	4 ft 0 in.	
5	501	603*									1 ft 9 in.
6	559	694	724								1 ft 10 in.
7	608	761	847								1 ft 11 in.
8	649	819	969								2 ft 0 in.
9	683	868	1,088								2 ft 0 in.
10	712	911	1,119	1,213							2 ft 1 in.
11	736	948	1,170	1,332							2 ft 2 in.
12	757	979	1,215	1,458							2 ft 3 in.
13	774	1,007	1,254	1,513	1,575						2 ft 4 in.
14	788	1,030	1,289	1,560	1,698						2 ft 4 in.
15	801	1,051	1,319	1,603	1,818						2 ft 5 in.
16	811	1,068	1,346	1,640	1,942						2 ft 6 in.
17	819	1,083	1,369	1,674	1,993	2,065					2 ft 7 in.
18	827	1,096	1,390	1,703	2,034	2,182					2 ft 7 in.
19	833	1,107	1,408	1,730	2,070	2,308					2 ft 8 in.
20	838	1,117	1,424	1,754	2,103	2,429					2 ft 9 in.
21	842	1,125	1,438	1,775	2,133	2,553					2 ft 9 in.
22	846	1,133	1,450	1,793	2,159	2,545	2,673				2 ft 10 in.
23	849	1,139	1,461	1,810	2,184	2,578	2,788				2 ft 11 in.
24	851	1,144	1,470	1,825	2,205	2,607	2,910				2 ft 11 in.
25	854	1,149	1,478	1,838	2,225	2,635	3,041				3 ft 0 in.

Table continued on next page

Backfill Loads in Pounds per Linear Foot on 8-in. Circular Pipe in a Trench Installation Based on 100 lb/ft³ Ordinary Clay Fill (continued)

Height of Backfill H Above Top of Pipe, ft	Trench Width at Top of Pipe										Transition Width†
	1 ft 6 in.	1 ft 9 in.	2 ft 0 in.	2 ft 3 in.	2 ft 6 in.	2 ft 9 in.	3 ft 0 in.	3 ft 3 in.	3 ft 6 in.	4 ft 0 in.	
26	855	1,153	1,486	1,850	2,242	2,659	**3,154**				3 ft 0 in.
27	857	1,156	1,492	1,861	2,258	2,682	3,128	**3,278**			3 ft 1 in.
28	858	1,159	1,498	1,870	2,273	2,702	3,155	**3,398**			3 ft 2 in.
29	859	1,162	1,502	1,878	2,286	2,721	3,181	**3,514**			3 ft 2 in.
30	860	1,164	1,507	1,886	2,297	2,738	3,204	**3,646**			3 ft 3 in.
31	861	1,166	1,511	1,892	2,308	2,753	3,225	**3,776**			3 ft 3 in.
32	862	1,167	1,514	1,898	2,317	2,767	3,245	3,748	**3,880**		3 ft 4 in.
33	862	1,169	1,517	1,904	2,326	2,780	3,263	3,772	**4,004**		3 ft 4 in.
34	862	1,170	1,519	1,908	2,333	2,791	3,279	3,794	**4,124**		3 ft 5 in.
35	863	1,171	1,522	1,913	2,340	2,802	3,294	3,815	**4,241**		3 ft 5 in.
36	863	1,172	1,524	1,916	2,346	2,811	3,308	3,834	**4,384**		3 ft 6 in.
37	863	1,173	1,525	1,920	2,352	2,820	3,321	3,851	**4,495**		3 ft 6 in.
38	864	1,173	1,527	1,922	2,357	2,828	3,333	3,868	4,431	**4,603**	3 ft 7 in.
39	864	1,174	1,528	1,925	2,362	2,835	3,343	3,883	4,451	**4,740**	3 ft 7 in.
40	864	1,174	1,529	1,927	2,366	2,842	3,353	3,896	4,470	**4,877**	3 ft 8 in.

Source: American Concrete Pipe Association, <www.concrete-pipe.org>.
* The bold printed figures are the maximum load at the transition width for any given height of backfill.
† The trench width at which the backfill fill load on the pipe is a maximum and remains constant regardless of increase in trench width.

Collection

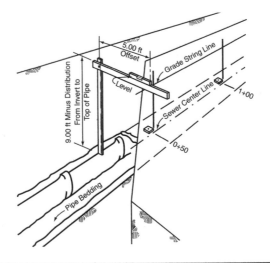

Grade Pole for Pipe Laying

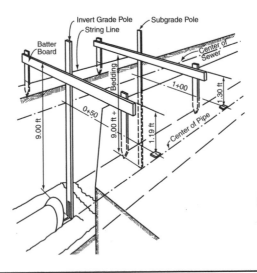

Grade Control Using Batter Boards

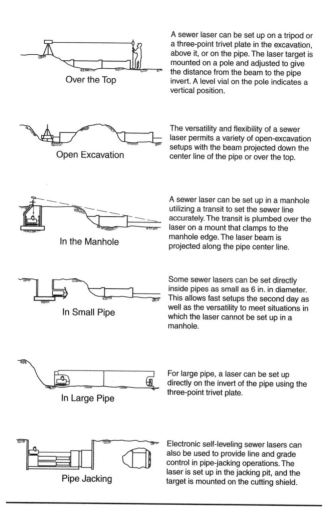

Over the Top

A sewer laser can be set up on a tripod or a three-point trivet plate in the excavation, above it, or on the pipe. The laser target is mounted on a pole and adjusted to give the distance from the beam to the pipe invert. A level vial on the pole indicates a vertical position.

Open Excavation

The versatility and flexibility of a sewer laser permits a variety of open-excavation setups with the beam projected down the center line of the pipe or over the top.

In the Manhole

A sewer laser can be set up in a manhole utilizing a transit to set the sewer line accurately. The transit is plumbed over the laser on a mount that clamps to the manhole edge. The laser beam is projected along the pipe center line.

In Small Pipe

Some sewer lasers can be set directly inside pipes as small as 6 in. in diameter. This allows fast setups the second day as well as the versatility to meet situations in which the laser cannot be set up in a manhole.

In Large Pipe

For large pipe, a laser can be set up directly on the invert of the pipe using the three-point trivet plate.

Pipe Jacking

Electronic self-leveling sewer lasers can also be used to provide line and grade control in pipe-jacking operations. The laser is set up in the jacking pit, and the target is mounted on the cutting shield.

Collection

Grade Control Using Fixed-Beam Laser

1. In lieu of a shoring system, the sides or walls of an excavation or trench may be sloped, provided equivalent protection is thus afforded. Where sloping is a substitute for shoring that would otherwise be needed, the slope shall be at least ¾ horizontal to 1 vertical unless the instability of the soil requires a slope flatter than ¾ to 1.

Exceptions: In hard, compact soil where the depth of the excavation or trench is 8 ft or less, a vertical cut of 3½ ft with sloping of ¾ horizontal to 1 vertical is permitted.

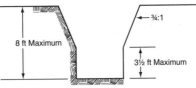

In hard, compact soil where the depth of the excavation or trench is 12 ft or less, a vertical cut of 3½ ft with sloping of 1 horizontal to 1 vertical is permitted.

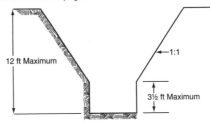

2. Benching in hard, compact soil is permitted provided that a slope ratio of ¾ horizontal to 1 vertical, or flatter, is used.

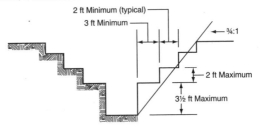

Sloping or Benching Systems

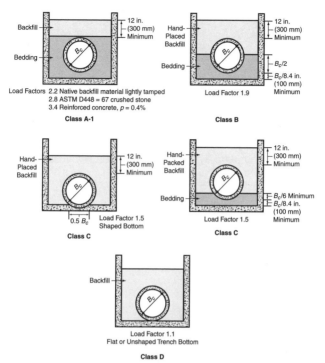

Class A-1

Load Factors 2.2 Native backfill material lightly tamped
2.8 ASTM D448 = 67 crushed stone
3.4 Reinforced concrete, $p = 0.4\%$

Class B

Load Factor 1.9

Class C

Load Factor 1.5
Shaped Bottom

Class C

Load Factor 1.5

Class D

Load Factor 1.1
Flat or Unshaped Trench Bottom

NOTE: The standard classes of rigid sewer-pipe bedding and their load factors (bedding factors) are shown. For example, an 8-in. vitrified clay pipe that has a three-edge bearing load supporting strength of 2,200 lb/ft will have a supporting strength of $(2,200 \text{ lb/ft} \times 1.5) = 3,300$ lb/ft when laid on a class C type of bedding.

Collection

Classes of Bedding

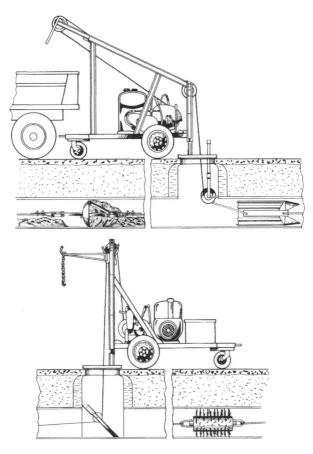

Courtesy of SRECO-Flexible, Inc.

Power Bucket Machines and Set Up

Highway Loads on Circular Pipe in Pounds per Linear Foot

Pipe Diameter, in.	Trench Width, ft	Height of Fill H Above Top of Pipe, ft												
		0.5	1.0	1.5	2.0	2.5	3.0	3.5	4.0	5.0	6.0	7.0	8.0	9.0
12	1.33	3,780	2,080	1,470	1,080	760	550	450	380	290	230	190	160	130
15	1.63	4,240	2,360	1,740	1,280	900	660	540	450	350	280	230	190	160
18	1.92	4,110	2,610	1,970	1,460	1,030	750	620	520	400	320	260	220	190
21	2.21	3,920	2,820	2,190	1,620	1,150	840	690	580	450	360	300	250	210
24	2.50	4,100	3,010	2,400	1,780	1,270	930	760	640	500	400	330	280	240
27	2.79	3,880	2,940	2,590	1,930	1,380	1,010	830	700	560	440	360	300	260
30	3.08	3,620	2,830	2,770	2,070	1,480	1,080	890	750	590	480	390	330	280
33	3.38	3,390	2,930	2,950	2,200	1,580	1,160	960	810	630	510	420	360	300
36	3.67	3,190	2,810	2,930	2,330	1,670	1,230	1,020	860	670	550	450	380	330
39	3.96	3,010	2,670	2,850	2,440	1,760	1,290	1,070	910	710	580	480	410	350
42	4.25	2,860	2,550	2,770	2,560	1,840	1,360	1,130	950	750	610	510	430	370
48	4.83	2,590	2,330	2,620	2,480	1,990	1,470	1,230	1,040	820	670	560	470	410
54	5.42	2,360	2,150	2,490	2,360	2,050	1,580	1,320	1,120	890	730	610	520	440
60	6.00	2,170	1,990	2,450	2,250	1,960	1,680	1,400	1,190	950	780	650	560	480
66	6.58	2,010	1,850	2,520	2,160	1,880	1,640	1,480	1,260	1,010	830	700	590	510
72	7.17	1,870	1,730	2,580	2,190	1,810	1,570	1,510	1,330	1,060	880	740	630	540
78	7.75	1,750	1,630	2,630	2,240	1,770	1,520	1,460	1,390	1,110	920	780	660	570

Table continued on next page

Collection

143

Highway Loads on Circular Pipe in Pounds per Linear Foot (continued)

Pipe Diameter, in.	Trench Width, ft	Height of Fill H Above Top of Pipe, ft												
		0.5	1.0	1.5	2.0	2.5	3.0	3.5	4.0	5.0	6.0	7.0	8.0	9.0
84	8.33	1,650	1,540	2,730	2,290	1,810	1,460	1,410	1,360	1,160	960	810	690	600
90	8.92	1,550	1,460	2,530	2,330	1,850	1,470	1,360	1,310	1,210	1,000	850	720	630
96	9.50	1,470	1,380	2,410	2,290	1,880	1,500	1,330	1,270	1,250	1,040	880	750	650
102	10.08	1,390	1,320	2,300	2,190	1,910	1,530	1,350	1,240	1,290	1,070	910	780	680
108	10.67	1,320	1,260	2,200	2,090	1,830	1,560	1,380	1,230	1,330	1,110	940	810	700
114	11.25	1,260	1,200	2,110	2,010	1,760	1,540	1,410	1,260	1,362	1,140	970	830	730
120	11.83	1,210	1,150	2,020	1,930	1,700	1,480	1,420	1,280	1,400	1,170	990	860	750
126	12.42	1,160	1,100	1,940	1,860	1,640	1,430	1,380	1,300	1,430	1,200	1,020	880	770
132	13.00	1,110	1,060	1,870	1,800	1,580	1,380	1,330	1,290	1,460	1,220	1,040	900	790
138	13.58	1,070	1,020	1,800	1,730	1,530	1,340	1,290	1,250	1,490	1,250	1,070	920	810
144	14.17	1,020	980	1,740	1,670	1,480	1,300	1,250	1,210	1,470	1,280	1,090	940	830

Source: American Concrete Pipe Association, <www.concrete-pipe.org>.

DATA: 1. Unsurfaced roadway.
2. Loads: American Association of State Highway and Transportation Officials HS 20, two 16,000-lb dual-tired wheels, 4 ft on centers; or alternate loading, four 12,000-lb dual-tired wheels, 4 ft on centers with impact included.

NOTES: 1. Interpolate for intermediate pipe sizes and/or fill heights.
2. Critical loads:
    a. For $H = 0.5$ and 1.0 ft, a single 16,000-lb dual-tired wheel.
    b. For $H = 1.5$–4.0 ft, two 16,000-lb dual-tired wheels, 4 ft on centers.
    c. For $H > 4.0$ ft, alternate loading.
3. Truck live loads for $H = 10.0$ ft or more are insignificant.

Recommended Impact Factors for Calculating Loads on Pipe With Less Than 3-ft Cover Subjected to Highway Truck Loads

Height of Cover *H*	Impact Factor
0 to 1 ft 0 in.	1.3
1 ft 1 in. to 2 ft 0 in.	1.2
2 ft 1 in. to 2 ft 11 in.	1.1
3 ft 0 in. and greater	1.0

Source: Standard Specifications for Highway Bridges, *by the American Association of State Highway and Transportation Officials, Washington, D.C. Used by permission.*

Crushing Strength Requirements for Vitrified Clay Sewer Pipe Based on the Three-Edge Bearing Test

Nominal Size, *in.*	Standard Strength, *lb/lin. ft*	Extra Strength, *lb/lin. ft*
4	1,200	2,000
6	1,200	2,000
8	1,400	2,200
10	1,600	2,400
12	1,800	2,600
15	2,000	2,900
18	2,200	3,300
21	2,400	3,850
24	2,600	4,400
27	2,800	4,700
30	3,300	5,000
33	3,600	5,500
36	4,000	6,000

Source: ASTM Specification C700, Standard and Extra Strength Clay Pipe. *Copyright ASTM INTERNATIONAL. Reprinted with permission.*

Collection

Strength Requirements for Reinforced Concrete Sewer Pipe Based on the Three-Edge Bearing Test (in pounds per linear foot of inside pipe diameter)

Classification	D Load to Produce 0.01-in. Crack	D Load at Failure	Pipe Size Diameter, *in.*
Class I	800	1,200	
Concrete strength 4,000 psi			60–96
Concrete strength 5,000 psi			102–108
Class II	1,000	1,500	
Concrete strength 4,000 psi			12–96
Concrete strength 5,000 psi			102–108
Class III	1,350	2,000	
Concrete strength 4,000 psi			12–72
Concrete strength 5,000 psi			78–108
Class IV	2,000	3,000	
Concrete strength 4,000 psi			12–66
Concrete strength 5,000 psi			60–84
Class V	3,000	3,750	
Concrete strength 6,000 psi			12–72

Source: ASTM Specification C76-66T. Copyright ASTM INTERNATIONAL. Reprinted with permission.

MANHOLES

Manholes provide an access to the sewer for inspection and maintenance operations. They also serve as ventilation, multiple pipe intersections, and pressure relief. Most manholes are cylindrical in shape.

The manhole cover must be secured so that it remains in place and avoids a blowout during peak flooding periods. Leakage from around the edges of the manhole cover should be kept to a minimum.

For small sewers, a minimum inside diameter of 4 ft (1.2 m) at the bottom tapering to a cast-iron frame that provides a clear opening usually specified as 2 ft (0.6 m) has been widely adopted. For sewers larger than 24 in. (600 mm), larger manhole bases are needed. Sometimes a platform is provided at one side, or the manhole is simply a vertical shaft over the center of the sewer.

Manholes are commonly located at the junctions of sanitary sewers, at changes in grades or alignment except in curved alignments, and at locations that provide ready access to the sewer for preventive maintenance and emergency service. Manholes are usually installed at street intersections.

Manhole spacing varies with available sanitary sewer maintenance methods. Typical manhole spacings range from 300 to 500 ft (90 to 150 m) in straight lines. For sewers larger than 5 ft (1.5 m), spacings of 500 to 1,000 ft (150 to 300 m) may be used.

Where the elevation difference between inflow and outflow sewers exceeds about 1.5 ft (0.5 m), sewer inflow that is dropped to the elevation of the outflow sewer by an inside or outside connection is called a drop manhole (or drop inlet). Its purpose is to protect workers from the splashing of wastewater, objectionable gases, and odors.

Collection

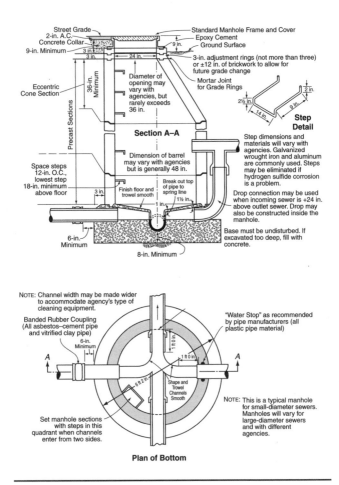

Street Grade
2-in. A.C.
Concrete Collar
9-in. Minimum
3 in.

Standard Manhole Frame and Cover
Epoxy Cement
Ground Surface
9 in.

3-in. adjustment rings (not more than three)
or ±12 in. of brickwork to allow for
future grade change

24 in.

Diameter of
opening may
vary with
agencies, but
rarely exceeds
36 in.

Mortar Joint
for Grade Rings

36-in.
Minimum

Eccentric
Cone Section

Precast Sections

Section A–A

Dimension of barrel
may vary with agencies,
but is generally 48 in.

2½ in.

14 in.

9 in.

2 in.

**Step
Detail**

Step dimensions and
materials will vary with
agencies. Galvanized
wrought iron and aluminum
are commonly used. Steps
may be eliminated if
hydrogen sulfide corrosion
is a problem.

Space steps
12-in. O.C.,
lowest step
18-in. minimum
above floor

3 in.

Finish floor and
trowel smooth

Break out top
of pipe to
spring line

1 in.

1½ in.

6-in.
Minimum

8-in. Minimum

Drop connection may be used
when incoming sewer is +24 in.
above outlet sewer. Drop may
also be constructed inside the
manhole.

Base must be undisturbed. If
excavated too deep, fill with
concrete.

NOTE: Channel width may be made wider
to accommodate agency's type of
cleaning equipment.

Banded Rubber Coupling
(All asbestos–cement pipe
and vitrified clay pipe)

6-in.
Minimum

"Water Stop" as recommended
by pipe manufacturers (all
plastic pipe material)

1 ft 0 in.

1 ft 0 in.

Shape and
Trowel
Channels
Smooth

A

A

Set manhole sections
with steps in this
quadrant when channels
enter from two sides.

NOTE: This is a typical manhole
for small-diameter sewers.
Manholes will vary for
large-diameter sewers
and with different
agencies.

Plan of Bottom

Precast Concrete Manhole

PIPE CHARACTERISTICS

Pressure Pipe

AWWA C900 refers to a category of standard dimension ratio (SDR) pipe that is the same diameter as ductile-iron (DI) pipe (ANSI/AWWA C900, *Polyvinyl Chloride (PVC) Pressure Pipe, and Fabricated Fittings, 4 in.–12 in. (100 mm–300 mm), for Water Distribution*). The following are all classified as C900 pipe. SDR/14 is pressure class 200, SDR/18 is pressure class 150, SDR/25 is pressure class 100.

The class signifies working pressure. SDR refers to a ratio of wall thickness to actual pipe outside diameter (OD). For example, SDR/18 pipe × 6.90 in. (the actual OD of 6-in. DI pipe) has a wall thickness of 6.90/18 = 0.38 in. Mechanical joints on C900 fittings are used with C900 pipe.

SDR/21 and SDR/26 have class designations that correspond to rated working pressure. The ratings incorporate a lower service factor than C900 pipe, which explains why SDR/21 and SDR/26 list a higher class rating for a given wall thickness. SDR/21 is class 200; SDR/26 is class 160.

The SDR numbers relate to wall thickness. SDR/21 × 6.63 in. (actual 6-in. steel pipe OD) has a wall thickness of 6.63/21 = 0.32 in.

| | Working Pressure, *psi* | | |
| Pipe Size, *in.* | Schedule 40 | Schedule 80 | |
	Socket	Socket	Threaded
1/2	600	850	420
3/4	480	690	340
1	450	630	320
1 1/4	370	520	260
1 1/2	330	471	240
2	300	425	210
2 1/2	280	400	200
3	260	375	190
4	220	324	160
6	180	280	140

Collection

Schedules 40 and 80 have the same diameter as steel pipe. The pressure ratings vary with the diameter of the pipe. The larger the diameter, the lower the rating.

SDR/21, SDR/26, and all Schedule pipe can be used with Schedule 40 and Schedule 80 fittings because they conform to steel pipe dimensions.

C900, SDR/21 and 26, and Schedule 40/80, can be used for sewer lines.

SDR/35 and SDR/41 are used exclusively for sewer drain only. Their outside dimensions are different from SDR pressure pipe and are different from each other in sizes other than 4 in. and 6 in.

Flange Guide

Gasket and Machine Bolt Dimensions for 150-lb Flange

Pipe Size, in.	Bolts Needed	Machine Bolt Dimension, in.	Gasket Dimensions	
			Ring, in.	Full Face, in.
2	4	$5/8 \times 2^{3}/4$	$2^{3}/8 \times 4^{1}/8$	$2^{3}/8 \times 6$
$2^{1}/2$	4	$5/8 \times 3$	$2^{7}/8 \times 4^{7}/8$	$2^{7}/8 \times 7$
3	4	$5/8 \times 3$	$3^{1}/2 \times 5^{3}/8$	$3^{1}/2 \times 7^{1}/2$
$3^{1}/2$	8	$5/8 \times 3$	$4 \times 6^{3}/8$	$4 \times 8^{1}/2$
4	8	$5/8 \times 3$	$4^{1}/2 \times 6^{7}/8$	$4^{1}/2 \times 9$
5	8	$3/4 \times 3^{1}/4$	$5^{9}/16 \times 7^{3}/4$	$5^{9}/16 \times 10$
6	8	$3/4 \times 3^{1}/4$	$6^{5}/8 \times 8^{3}/4$	$6^{5}/8 \times 11$
8	8	$3/4 \times 3^{1}/2$	$8^{5}/8 \times 11$	$8^{5}/8 \times 13^{1}/2$
10	12	$7/8 \times 3^{3}/4$	$10^{3}/4 \times 13^{3}/8$	$10^{3}/4 \times 16$
12	12	$7/8 \times 4$	$12^{3}/4 \times 16^{1}/8$	$12^{3}/4 \times 19$

Types of Plastic Pipe

Although there are many types of plastic pipe, PVC and ABS are by far the most common. It is very important that the correct primers and solvents be used on each type of pipe or the joints will not seal properly and the overall strength will be weakened.

Types	Characteristics
ABS (acrilonitrile butadiene styrene)	Strong, rigid, and resistant to a variety of acids and bases. Some solvents and chlorinated hydrocarbons may damage the pipe. Maximum usable temperature is 160°F (71°C) at low pressures. It is most common as a drain, waste, and vent pipe.
CPVC (chlorinated polyvinyl chloride)	Similar to PVC but designed specifically for piping water at up to 180°F (82°C). Pressure rating is 100 psi.
FRP (fiberglass-reinforced plastic) epoxy	A thermosetting plastic over fiberglass. Very high strength and excellent chemical resistance. Good to 220°F (105°C). Excellent for labs.
PB (polybutylene)	A flexible pipe for pressurized water systems, both hot and cold. *Only* compression and banded-type joints can be used.
PE (polyethylene)	A flexible pipe for pressurized water systems such as sprinklers. Not for hot water.
Polypropylene	Low pressure, lightweight material that is good up to 180°F (82°C). Highly resistant to acids, bases, and many solvents. Good for laboratory plumbing.
PVC (polyvinyl chloride)	Strong, rigid, and resistant to a variety of acids and bases. Some solvents and chlorinated hydrocarbons may damage the pipe. Can be used with water, gas, and drainage systems but *not* with hot-water systems.
PVDF (polyvinylidene fluoride)	Strong, very tough, and resistant to abrasions, acids, bases, solvents, and much more. Good to 280°F (138°C). Good in lab.

Collection

Outside Diameter of Small Pipe and Tube

Type of Pipe or Tubing	Nominal Pipe Size, *in.*								
	1/4	3/8	1/2	5/8	3/4	1	1 1/4	1 1/2	2
Copper and CTS-PE*	.375	.500	.625	.750	.875	1.125	1.375	1.625	2.125
Iron pipe and IPS-PE†	.540	.675	.840	—	1.050	1.315	1.660	1.900	2.375
Lead pipe‡									
Strong	—	—	—	1.010	1.156	1.428	—	—	—
Extra strong	—	—	.876	1.082	1.212	1.492	1.765	2.076	2.751
Double extra strong	—	—	1.012	1.335	1.596	—	—	—	—

NOTE: Polyethylene (PE) pipe is also available with the same inside diameter as iron pipe. The wall thickness varies with the pressure class, so the outside diameter (OD) is variable. See manufacturer's information or ANSI/AWWA Standard C901 for details.

  * CTS-PE—Copper tubing–size polyethylene. Tubing has the same OD as copper tube.
  † IPS-PE—Iron pipe–size polyethylene. Pipe has the same OD as iron pipe.
  ‡ The OD of lead pipe is approximate.

Smoothness Coefficients for Various Pipe Materials

Type of Pipe	*C* Value	
	New	10 Years Old
Asbestos–cement pipe	140	120–130
Cast-iron pipe	130	100
Reinforced and plain concrete pipe	140	120–130
Ductile-iron pipe	130	100
Plastic pipe	150	120–130
Steel pipe	110	100

Friction Loss Factors for 12-in. Pipe

Discharge		Velocity,	Velocity Head,	Loss of Head, ft per 1,000 ft of length						
gpd	ft³/sec	ft/sec	ft	C = 140	C = 130	C = 120	C = 110	C = 100	C = 90	C = 80
100,000	0.155	0.20	0.00	0.02	0.02	0.02	0.02	0.03	0.04	0.04
200,000	0.309	0.39	0.00	0.06	0.07	0.08	0.09	0.11	0.13	0.16
300,000	0.464	0.59	0.01	0.12	0.14	0.16	0.19	0.22	0.27	0.34
400,000	0.619	0.79	0.01	0.20	0.24	0.27	0.32	0.38	0.47	0.58
500,000	0.774	0.99	0.02	0.31	0.36	0.41	0.48	0.58	0.71	0.88
600,000	0.928	1.18	0.02	0.44	0.50	0.58	0.68	0.81	0.99	1.23
700,000	1.083	1.38	0.03	0.58	0.66	0.77	0.91	1.08	1.32	1.64
800,000	1.238	1.58	0.04	0.74	0.85	0.99	1.15	1.38	1.68	2.09
900,000	1.392	1.77	0.05	0.92	1.06	1.23	1.45	1.72	2.10	2.61
1,000,000	1.547	1.97	0.06	1.12	1.29	1.50	1.76	2.10	2.57	3.18
1,100,000	1.702	2.17	0.07	1.34	1.54	1.79	2.10	2.50	3.04	3.79
1,200,000	1.857	2.36	0.09	1.58	1.81	2.10	2.47	2.94	3.58	4.45
1,300,000	2.011	2.56	0.10	1.83	2.10	2.43	2.85	3.40	4.14	5.2
1,400,000	2.166	2.76	0.12	2.10	2.40	2.79	3.26	3.90	4.76	5.9
1,500,000	2.321	2.96	0.14	2.39	2.73	3.17	3.71	4.43	5.4	6.7
1,600,000	2.476	3.15	0.15	2.69	3.09	3.58	4.20	5.0	6.1	7.6
1,700,000	2.630	3.35	0.17	3.00	3.45	4.00	4.69	5.6	6.8	8.5
1,800,000	2.785	3.55	0.20	3.33	3.82	4.43	5.2	6.2	7.6	9.4
1,900,000	2.940	3.74	0.22	3.70	4.24	4.92	5.8	6.9	8.4	10.4
2,000,000	3.094	3.94	0.24	4.06	4.65	5.4	6.4	7.6	9.2	11.5

Table continued on next page

Collection

153

Friction Loss Factors for 12-in. Pipe (continued)

Discharge		Velocity, ft/sec	Velocity Head, ft	Loss of Head, ft per 1,000 ft of length							
gpd	ft³/sec			C = 140	C = 130	C = 120	C = 110	C = 100	C = 90	C = 80
2,000,000	3.094	3.94	0.24	4.06	4.65	5.4	6.4	7.6	9.2	11.5
2,200,000	3.404	4.33	0.29	4.85	5.6	6.5	7.6	9.0	10.9	13.7
2,400,000	3.713	4.73	0.35	5.7	6.5	7.6	8.9	10.5	12.8	16.0
2,600,000	4.023	5.12	0.41	6.6	7.6	8.8	10.3	12.3	15.0	18.6
2,800,000	4.332	5.52	0.47	7.6	8.7	10.1	11.9	14.1	17.2	21.5
3,000,000	4.642	5.91	0.54	8.6	9.9	11.5	13.5	16.0	19.4	24.3
3,500,000	5.41	6.89	0.74	11.4	13.2	15.3	17.9	21.3	26.0	32.3
4,000,000	6.19	7.88	0.96	14.5	16.6	19.3	22.6	27.0	33.2	41
4,500,000	6.96	8.87	1.22	18.0	20.6	24.0	28.2	33.6	41.2	51
5,000,000	7.74	9.85	1.50	22.0	25.1	29.2	34.3	41.0	50.0	62
5,500,000	8.51	10.84	1.82	26.5	30.3	35.1	41.4	49.4	60	75
6,000,000	9.28	11.82	2.17	31.1	35.7	41.4	48.8	58	70	88
7,000,000	10.83	13.79	2.96	41.2	47.2	55	65	77	94	116
8,000,000	12.38	15.76	3.86	53	61	71	83	99	121	150
9,000,000	13.92	17.73	4.89	66	75	87	103	122	148	185
10,000,000	15.47	19.70	6.03	81	93	107	126	150	183	228

Friction Loss of Water, in Feet per 100-ft Length of Pipe, Based on Hazen–Williams Formula for C = 100

gpm	½-in Vel, ft/sec	½-in Loss, ft	¾-in Vel, ft/sec	¾-in Loss, ft	1-in Vel, ft/sec	1-in Loss, ft	1¼-in Vel, ft/sec	1¼-in Loss, ft	1½-in Vel, ft/sec	1½-in Loss, ft	2-in Vel, ft/sec	2-in Loss, ft	2½-in Vel, ft/sec	2½-in Loss, ft	3-in Vel, ft/sec	3-in Loss, ft	4-in Vel, ft/sec	4-in Loss, ft	5-in Vel, ft/sec	5-in Loss, ft	6-in Vel, ft/sec	6-in Loss, ft
2	2.10	7.4	1.20	1.9																		
4	4.21	27.0	2.41	7.0	1.49	2.14	.86	.57	.63	.26												
6	6.31	57.0	3.61	14.7	2.23	4.55	1.29	1.20	.94	.56	.61	.20										
8	8.42	98.0	4.81	25.0	2.98	7.8	1.72	2.03	1.26	.95	.82	.33	.52	.11								
10	10.52	147.0	6.02	38.0	3.72	11.7	2.14	3.05	1.57	1.43	1.02	.50	.65	.17	.45	.07						
12			7.22	53.0	4.46	16.4	2.57	4.3	1.89	2.01	1.23	.79	.78	.23	.54	.10						
15			9.02	80.0	5.60	25.0	3.21	6.5	2.36	3.00	1.53	1.08	.98	.36	.68	.15						
18			10.84	108.2	6.69	35.0	3.86	9.1	2.83	4.24	1.84	1.49	1.18	.50	.82	.21						
20			12.03	136.0	7.44	42.0	4.29	11.1	3.15	5.20	2.04	1.82	1.31	.61	.91	.25	.51	.06				
25					9.30	64.0	5.36	16.6	3.80	7.30	2.55	2.73	1.63	.92	1.13	.38	.64	.09				
30					11.15	89.0	6.43	23.0	4.72	11.0	3.06	3.84	1.96	1.29	1.36	.54	.77	.13	.49	.04		
35					13.02	119.0	7.51	31.2	5.51	14.7	3.57	5.10	2.29	1.72	1.59	.71	.89	.17	.57	.06		
40					14.88	152.0	8.58	40.0	6.30	18.8	4.08	6.6	2.61	2.20	1.82	.91	1.02	.22	.65	.08		
45							9.65	50.0	7.08	23.2	4.60	8.2	2.94	2.80	2.04	1.15	1.15	.28	.73	.09		
50							10.72	60.0	7.87	28.4	5.11	9.9	3.27	3.32	2.27	1.38	1.28	.34	.82	.11	.57	.04
55							11.78	72.0	8.66	34.0	5.62	11.8	3.59	4.01	2.45	1.58	1.41	.41	.90	.14	.62	.05
60							12.87	85.0	9.44	39.6	6.13	13.9	3.92	4.65	2.72	1.92	1.53	.47	.98	.16	.68	.06
65							13.92	99.7	10.23	45.9	6.64	16.1	4.24	5.4	2.89	2.16	1.66	.53	1.06	.19	.74	.076

Table continued on next page

Collection

155

Friction Loss of Water, in Feet per 100-ft Length of Pipe, Based on Hazen–Williams Formula for C = 100 (continued)

gpm	½-in. Pipe Vel., ft/sec	Loss, ft	¾-in. Pipe Vel., ft/sec	Loss, ft	1-in. Pipe Vel., ft/sec	Loss, ft	1¼-in. Pipe Vel., ft/sec	Loss, ft	1½-in. Pipe Vel., ft/sec	Loss, ft	2-in. Pipe Vel., ft/sec	Loss, ft	2½-in. Pipe Vel., ft/sec	Loss, ft	3-in. Pipe Vel., ft/sec	Loss, ft	4-in. Pipe Vel., ft/sec	Loss, ft	5-in. Pipe Vel., ft/sec	Loss, ft	6-in. Pipe Vel., ft/sec	Loss, ft
70							15.01	113.0	11.02	53.0	7.15	18.4	4.58	6.2	3.18	2.57	1.79	.63	1.14	.21	.79	.08
75							16.06	129.0	11.80	60.0	7.66	20.9	4.91	7.1	3.33	3.00	1.91	.73	1.22	.24	.85	.10
80							17.16	145.0	12.59	68.0	8.17	23.7	5.23	7.9	3.63	3.28	2.04	.81	1.31	.27	.91	.11
85							18.21	163.8	13.38	75.0	8.68	26.5	5.56	8.1	3.78	3.54	2.17	.91	1.39	.31	.96	.12
90							19.30	180.0	14.71	84.0	9.19	29.4	5.88	9.8	4.09	4.08	2.30	1.00	1.47	.34	1.02	.14
95									14.95	93.0	9.70	32.6	6.21	10.8	4.22	4.33	2.42	1.12	1.55	.38	1.08	.15
100									15.74	102.0	10.21	35.8	6.54	12.0	4.54	4.96	2.55	1.22	1.63	.41	1.13	.17
110									17.31	122.0	11.23	42.9	7.18	14.5	5.00	6.0	2.81	1.46	1.79	.49	1.25	.21
120									18.89	143.0	12.25	50.0	7.84	16.8	5.45	7.0	3.06	1.17	1.96	.58	1.36	.24
130									20.46	166.0	13.28	58.0	8.48	18.7	5.91	8.1	3.31	1.97	2.12	.67	1.47	.27
140									22.04	190.0	14.30	67.0	9.15	22.3	6.35	9.2	3.57	2.28	2.29	.76	1.59	.32
150											15.32	76.0	9.81	25.5	6.82	10.5	3.82	2.62	2.45	.88	1.70	.36
160											16.34	86.0	10.46	29.0	7.26	11.8	4.08	2.91	2.61	.98	1.82	.40
170											17.36	96.0	11.11	34.1	7.71	13.3	4.33	3.26	2.77	1.08	1.92	.45
180											18.38	107.0	11.76	35.7	8.17	14.0	4.60	3.61	2.94	1.22	2.04	.50
190											19.40	118.0	12.42	39.6	8.63	15.5	4.84	4.01	3.10	1.35	2.16	.55
200											20.42	129.0	13.07	43.1	9.08	17.8	5.11	4.4	3.27	1.48	2.27	.62
220											22.47	154.0	14.38	52.0	9.99	21.3	5.62	5.2	3.59	1.77	2.50	.73

8-in. Pipe

gpm	Vel., ft/sec	Loss, ft
140	.90	.08
150	.96	.09
160	1.02	.10
170	1.08	.11
180	1.15	.13
190	1.21	.14
200	1.28	.15
220	1.40	.18

10-in. Pipe

gpm	Vel., ft/sec	Loss, ft
220	.90	.06

Table continued on next page

156

Friction Loss of Water, in Feet per 100-ft Length of Pipe, Based on Hazen–Williams Formula for $C = 100$ (continued)

Note: The columns printed as "1-in. Pipe," "1¼-in. Pipe," and "1½-in. Pipe" carry boxed inset data for **12-in. Pipe**, **14-in. Pipe**, and **16-in. Pipe**, respectively (the small pipes cannot carry these flows).

gpm	8-in. Pipe Vel., ft/sec	8-in. Pipe Loss, ft	10-in. Pipe Vel., ft/sec	10-in. Pipe Loss, ft	1-in. Pipe (12-in. inset) Vel., ft/sec	Loss, ft	1¼-in. Pipe (14-in. inset) Vel., ft/sec	Loss, ft	1½-in. Pipe (16-in. inset) Vel., ft/sec	Loss, ft	2-in. Pipe Vel., ft/sec	Loss, ft	2½-in. Pipe Vel., ft/sec	Loss, ft	3-in. Pipe Vel., ft/sec	Loss, ft	4-in. Pipe Vel., ft/sec	Loss, ft	5-in. Pipe Vel., ft/sec	Loss, ft	6-in. Pipe Vel., ft/sec	Loss, ft
240	1.53	.22	.98	.07							24.51	182.0	15.69	61.0	10.89	25.1	6.13	6.2	3.92	2.08	2.72	.87
260	1.66	.25	1.06	.08							26.55	211.0	16.99	70.0	11.80	29.1	6.64	7.2	4.25	2.41	2.95	1.00
280	1.79	.28	1.15	.09									18.30	81.0	12.71	33.4	7.15	8.2	4.58	2.77	3.18	1.14
300	1.91	.32	1.22	.11									19.61	92.0	13.62	38.0	7.66	9.3	4.90	3.14	3.40	1.32
320	2.05	.37	1.31	.12									20.92	103.0	14.52	42.8	8.17	10.5	5.23	3.54	3.64	1.47
340	2.18	.41	1.39	.14									22.22	116.0	15.43	47.9	8.68	11.7	5.54	3.97	3.84	1.62
360	2.30	.45	1.47	.15									23.53	128.0	16.34	53.0	9.19	13.1	5.87	4.41	4.08	1.83
380	2.43	.50	1.55	.17	1.08	.069							24.84	142.0	17.25	59.0	9.69	14.0	6.19	4.86	4.31	2.00
400	2.60	.54	1.63	.19	1.14	.075							26.14	156.0	18.16	65.0	10.21	16.0	6.54	5.4	4.55	2.20
450	2.92	.68	1.84	.23	1.28	.095									20.40	78.0	11.49	19.8	7.35	6.7	5.11	2.74
500	3.19	.82	2.04	.28	1.42	.113	1.04	.06							22.70	98.0	12.77	24.0	8.17	8.1	5.68	2.90
550	3.52	.97	2.24	.33	1.56	.135	1.15	.07							24.96	117.0	14.04	28.7	8.99	9.6	6.25	3.96
600	3.84	1.14	2.45	.39	1.70	.159	1.25	.08							27.23	137.0	15.32	33.7	9.80	11.3	6.81	4.65
650	4.16	1.34	2.65	.45	1.84	.19	1.37	.09									16.59	39.0	10.62	13.2	7.38	5.40
700	4.46	1.54	2.86	.52	1.99	.22	1.46	.10									17.87	44.9	11.44	15.1	7.95	6.21
750	4.80	1.74	3.06	.59	2.13	.24	1.58	.11									19.15	51.0	12.26	17.2	8.50	7.12
800	5.10	1.90	3.26	.66	2.27	.27	1.67	.13									20.42	57.0	13.07	19.4	9.08	7.96
850	5.48	2.20	3.47	.75	2.41	.31	1.79	.14	1.36	.08							21.70	64.0	13.89	21.7	9.65	8.95

Table continued on next page

Collection

157

Friction Loss of Water, in Feet per 100-ft Length of Pipe, Based on Hazen–Williams Formula for C = 100 (continued)

gpm	8-in. Pipe		10-in. Pipe		12-in. Pipe		14-in. Pipe		16-in. Pipe		2-in. Pipe		2½-in. Pipe		3-in. Pipe		4-in. Pipe		5-in. Pipe		6-in. Pipe	
	Vel., ft/sec	Loss, ft	Vel., ft/sec	Loss, ft	Vel., ft/sec	Loss, ft	Vel., ft/sec	Loss, ft	Vel., ft/sec	Loss, ft	Vel., ft/sec	Loss, ft	Vel., ft/sec	Loss, ft	Vel., ft/sec	Loss, ft	Vel., ft/sec	Loss, ft	Vel., ft/sec	Loss, ft	Vel., ft/sec	Loss, ft
900	5.75	2.46	3.67	.83	2.56	.34	1.88	.16	1.44	.084							22.98	71.0	14.71	24.0	10.20	10.11
950	6.06	2.87	3.88	.91	2.70	.38	2.00	.18	1.52	.095	20-in. Pipe								15.52	26.7	10.77	11.20
1,000	6.38	2.97	4.08	1.03	2.84	.41	2.10	.19	1.60	.10	1.02	.04							16.34	29.2	11.34	12.04
1,100	7.03	3.52	4.49	1.19	3.13	.49	2.31	.23	1.76	.12	1.12	.04							17.97	34.9	12.48	14.55
1,200	7.66	4.17	4.90	1.40	3.41	.58	2.52	.27	1.92	.14	1.23	.05							19.61	40.9	13.61	17.10
1,300	8.30	4.85	5.31	1.62	3.69	.67	2.71	.32	2.08	.17	1.33	.06									14.72	18.4
1,400	8.95	5.50	5.71	1.87	3.98	.78	2.92	.36	2.24	.19	1.43	.064									15.90	22.60
1,500	9.58	6.24	6.12	2.13	4.26	.89	3.15	.41	2.39	.21	1.53	.07									17.02	25.60
1,600	10.21	7.00	6.53	2.39	4.55	.98	3.34	.47	2.56	.24	1.63	.08	24-in. Pipe								18.10	26.9
1,800	11.50	8.78	7.35	2.95	5.11	1.21	3.75	.58	2.87	.30	1.84	.10	1.28	.04								
2,000	12.78	10.71	8.16	3.59	5.68	1.49	4.17	.71	3.19	.37	2.04	.12	1.42	.05								
2,200	14.05	12.78	8.98	4.24	6.25	1.81	4.59	.84	3.51	.44	2.25	.15	1.56	.06	30-in. Pipe							
2,400	15.32	14.2	9.80	5.04	6.81	2.08	5.00	.99	3.83	.52	2.45	.17	1.70	.07	1.09	.02						
2,600			10.61	5.81	7.38	2.43	5.47	1.17	4.15	.60	2.66	.20	1.84	.08	1.16	.027						
2,800			11.41	6.70	7.95	2.75	5.84	1.32	4.47	.68	2.86	.23	1.98	.09	1.27	.03						
3,000			12.24	7.62	8.52	3.15	6.01	1.49	4.79	.78	3.08	.27	2.13	.10	1.37	.037						
3,200			13.05	7.8	9.10	3.51	6.68	1.67	5.12	.88	3.27	.30	2.26	.12	1.46	.041						
3,500			14.30	10.08	9.95	4.16	7.30	1.97	5.59	1.04	3.59	.35	2.49	.14	1.56	.047						

Table continued on next page

Friction Loss of Water, in Feet per 100-ft Length of Pipe, Based on Hazen–Williams Formula for C = 100 (continued)

	8-in. Pipe		10-in. Pipe		12-in. Pipe		14-in. Pipe		16-in. Pipe		20-in. Pipe		24-in. Pipe		30-in. Pipe		4-in. Pipe		5-in. Pipe		6-in. Pipe	
gpm	Vel., ft/sec	Loss, ft	Vel., ft/sec	Loss, ft	Vel., ft/sec	Loss, ft	Vel., ft/sec	Loss, ft	Vel., ft/sec	Loss, ft	Vel., ft/sec	Loss, ft	Vel., ft/sec	Loss, ft	Vel., ft/sec	Loss, ft	Vel., ft/sec	Loss, ft	Vel., ft/sec	Loss, ft	Vel., ft/sec	Loss, ft
3,800			15.51	13.4	10.80	4.90	7.98	2.36	6.07	1.20	3.88	.41	2.69	.17	1.73	.05						
4,200					11.92	5.88	8.76	2.77	6.70	1.44	4.29	.49	2.99	.20	1.91	.07						
4,500					12.78	6.90	9.45	3.22	7.18	1.64	4.60	.56	3.20	.22	2.04	.08						
5,000					14.20	8.40	10.50	3.92	8.01	2.03	5.13	.68	3.54	.27	2.26	.09						
5,500							11.55	4.65	8.78	2.39	5.64	.82	3.90	.33	2.50	.11						
6,000							12.60	5.50	9.58	2.79	6.13	.94	4.25	.38	2.73	.13						
6,500							13.65	6.45	10.39	3.32	6.64	1.10	4.61	.45	2.96	.15						
7,000							14.60	7.08	11.18	3.70	7.15	1.25	4.97	.52	3.18	.17						
8,000									12.78	4.74	8.17	1.61	5.68	.66	3.64	.23						
9,000									14.37	5.90	9.20	2.01	6.35	.81	4.08	.28						
10,000									15.96	7.19	10.20	2.44	7.07	.98	4.54	.33						
12,000											12.25	3.41	8.50	1.40	5.46	.48						
14,000											14.30	4.54	9.95	1.87	6.37	.63						
16,000													11.38	2.40	7.28	.81						
18,000													12.76	2.97	8.18	1.02						
20,000													14.20	3.60	9.10	1.23						

Collection

PIPE JOINTS

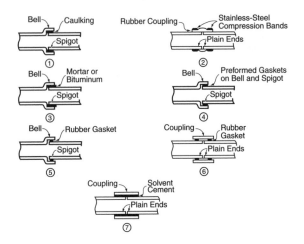

Joint No.	Joint Description	Joint Used With*							
		ABSP	ACP	CIP	PVCP	RCP	CP	RPMP	VCP
1	Caulked Bell and Spigot			✓					
2	Coupling for Plain-End Pipe			✓					✓
3	Mortar/Bituminous–Filled Bell and Spigot					✓	✓		✓
4	Polyvinyl Chloride/Polyurethane Preformed Gaskets in Bell and Spigot								✓
5	Rubber-Gasketed Ball and Spigot			✓	✓	✓		✓	✓
6	Rubber-Gasketed Coupling		✓						
7	Solvent-Cemented Coupling	✓							

*ABSP = acrylonitrile–butadiene–styrene pipe; ACP = asbestos–cement pipe;
CIP = cast-iron pipe; CP = concrete pipe; PVCP = polyvinyl chloride pipe;
RCP = reinforced concrete pipe; RPMP = reinforced plastic mortar pipe;
VCP = vitrified clay pipe.

Common Types of Joints for Sewer Pipe

Types of Joint Breaks

Drop Joint
Pulling video camera from right to left past this joint will cause it to drop.

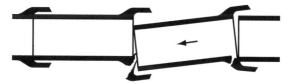

Jump Joint
Video camera must be "jumped" from right to left to pass this type
of situation.

Pulling the camera past situations like these can prevent the camera from
being pulled back out if a full obstruction is encountered later down the pipe.
It is best to document conditions like these, repair them, and then reinspect
the line.

Drop Joint and Jump Joint

Backflow Preventers

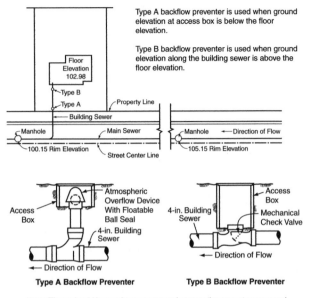

Type A backflow preventer is used when ground elevation at access box is below the floor elevation.

Type B backflow preventer is used when ground elevation along the building sewer is above the floor elevation.

Floor Elevation 102.98

Type B
Type A
Property Line
Building Sewer
Manhole
Main Sewer
Manhole ← Direction of Flow
100.15 Rim Elevation
Street Center Line
105.15 Rim Elevation

Atmospheric Overflow Device With Floatable Ball Seal
Access Box
4-in. Building Sewer
← Direction of Flow

Type A Backflow Preventer

Access Box
4-in. Building Sewer
Mechanical Check Valve
← Direction of Flow

Type B Backflow Preventer

NOTE: There should be a written agreement between the property owner and the sewer district regarding who is responsible for the maintenance of backflow preventers. Grease deposits or rags can prevent their proper operation, and they should be cleaned every 6 months.

Location and Type of Wastewater Backflow Preventers

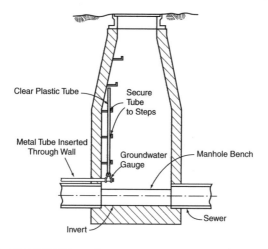

Clear Plastic Tube

Secure Tube to Steps

Metal Tube Inserted Through Wall

Groundwater Gauge

Manhole Bench

Sewer

Invert

Typical Static Groundwater Gauge Installation

Collection

Open

Eccentric action and resilient seating assure lasting dead-tight shutoff. As the eccentric plug rotates 90° from open to closed, it moves into a raised eccentric seat.

In the open position, flow is straight through and flow capacity is high.

Closing

As the plug closes, it moves toward the seat without scraping the seat or body walls so there is no plug binding or wear. Flow is still straight through making the throttling characteristic of this valve ideal for manual throttling of gases and liquids.

Closed

In the closed position, the plug makes contact with the seat. The resilient plug seal is pressed firmly into the seat for dead-tight shutoff. Eccentric plug and seat design ensure lasting shutoff because the plug continues to move into the seat until firm contact and seal is made.

Courtesy of DeZUIRK Water Controls.

Open, Closing, and Closed Eccentric Valve

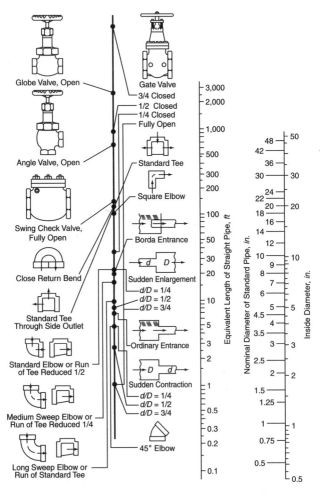

Courtesy of Crane Valves–North America.

Resistance of Valves and Fittings to Flow of Fluids

TYPES OF CORROSION

There are many ways to classify corrosion, but it is mostly based on its forms and causes.

1. *Physical corrosion:* This is erosion due to high-velocity (i.e., more than 5 ft/sec) particulate matter or when dispersed gas bubbles erode the surface of a pipe.

2. *Stray current corrosion:* This occurs as a localized attack caused by the potential differential from an outside source, such as grounding appliances, through pipes. Corrosion occurs where the current leaves the pipe.

3. *Uniform corrosion:* This takes place at an equal rate over all metal the surface due to the uniform presence of many electrochemical cells throughout the surface. This is mostly associated with low pH of the water. Any one site on the metal surface may be anodic one instant and cathodic the next. The loss in weight is directly proportional to the time of exposure.

4. *Localized or pitting corrosion:* This occurs where there is an exposed area of an otherwise coated metal surface due to stress. This localized area becomes anodic and is surrounded by a large cathodic area. The pit remains anodic and becomes larger and larger as corrosion progresses. Often, the pit is covered with a tubercle.

5. *Concentration cell corrosion:* This is a function of the difference in concentration of various dissolved substances, such as metal and hydrogen ions or oxygen molecules, on the adjacent portions of the metal surface. The corrosion process will tend to equalize the metal ion concentration by dissolving the areas with less concentration, which become anodic. Similarly, the part with less oxygen acts anodic and the one with more oxygen acts cathodic.

6. *Galvanic corrosion:* This occurs when two dissimilar metals are connected together in the water lines. One metal becomes anodic and the other cathodic. A metal on top (more active) in an electromotive series becomes an anode and the one below, the cathode. For example, lead-solder

Collection

and copper pipe will have lead as the anode and copper as the cathode, causing the corrosion of lead.

7. *Bacterial corrosion:* At dead ends of pipes where anaerobic conditions exist (lack of oxygen), sulfates are converted into hydrogen sulfide (H_2S) by anaerobic bacteria. Because H_2S is an acid, it increases H^+ concentration and dissolves iron, which causes very unpleasant tastes and odors.

VARIOUS FACTORS AFFECTING CORROSION

Dissolved Solids

Different types of ions and their concentrations have different effects on corrosion. Carbonates, polyphosphates, and silicates normally reduce corrosion by forming a protective film on the metal surface whereas chlorides may increase corrosion by creating acidic conditions that interfere with the protective film.

Dissolved Gases

Carbon dioxide and oxygen are the most common dissolved gases. Carbon dioxide forms carbonic acid (H_2CO_3) in water. It is an amphiprotic compound with an important buffering effect in water. Oxygen acts cathodic and as a depolarizer as it combines with hydrogen to form water. A higher concentration of oxygen causes more corrosion, and vice versa. Other gases of interest are chlorine, ammonia, and hydrogen sulfide. Chlorine as hypochlorous acid causes acidic conditions and thus more corrosion. Hydrogen sulfide is also an acid and thus causes corrosion. Ammonia forms chloramines with chlorine, which are known to inhibit corrosion.

Temperature

As a rule, the higher the temperature, the higher the rate of corrosion. There is less corrosion during winter than summer. Soft waters (with alkalinity less than 50 mg/L) are very corrosive during summer.

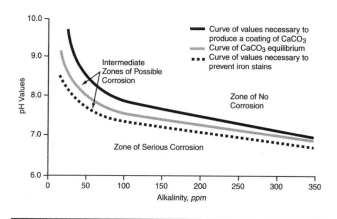

Stability Curve Showing Relationship Between pH Values and Alkalinity

Various Corrosion Indices

1. *Marble test:* This test is based on the alkalinity of the water. The water to be tested is saturated with calcium carbonate ($CaCO_3$) by adding $CaCO_3$ powder to it, shaking it, and keeping it overnight. Alkalinity of the sample is determined before and after saturation. If the initial alkalinity is equal to the final alkalinity, the water is stable. If the initial alkalinity is higher, then water is depositing; if it is less, then it is corrosive. This is a simple, easy, and good way for an operator to control corrosion.

2. *Baylis curve:* This shows the solubility of $CaCO_3$ with regard to alkalinity and pH (see figure above). If the point of intersection of pH and alkalinity of the water is above the equilibrium curve, the water is depositing; if it is below, the water is corrosive; and if it is equal, the water is stable.

3. *Langelier saturation index (LSI):* This is a commonly used index in the water utility industry. It determines the $CaCO_3$ deposition property of the water by calculating saturation pH (pH_s). pH_s is calculated from total dissolved solids, temperature, alkalinity, and calcium content of the water. If

the pH of the water is equal to the pH_s, the water is stable; if it is higher, the water is depositing; if it is less, the water is corrosive.

$$LSI = pH - pH_s$$

4. *Ryzner index (RI) or stability index:* This index is two times pH_s minus pH of the water.

$$RI = 2pH_s - pH$$

An RI value less than 6 indicates a depositing water, and more than 6 indicates corrosion. The higher the RI, the more corrosive the water in the lines.

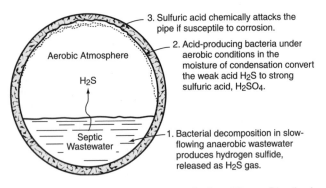

3. Sulfuric acid chemically attacks the pipe if susceptile to corrosion.

2. Acid-producing bacteria under aerobic conditions in the moisture of condensation convert the weak acid H_2S to strong sulfuric acid, H_2SO_4.

1. Bacterial decomposition in slow-flowing anaerobic wastewater produces hydrogen sulfide, released as H_2S gas.

Aerobic Atmosphere

H_2S

Septic Wastewater

Courtesy of Pearson Education, Inc.

Environmental Conditions Leading to Crown Corrosion Occur in Sanitary Sewers as a Result of Flat Grades and Warm Wastewaters

PIPE TESTING

Minimum Test Times per 100 ft of Sewer Line for Low-Pressure Air Testing

Pipe Diameter, *in.*	Test Time, *min*	Pipe Diameter, *in.*	Test Time, *min*
3	0.2	21	3.0
4	0.3	24	3.6
6	0.7	27	4.2
8	1.2	30	4.8
10	1.5	33	5.4
12	1.8	36	6.0
15	2.1	39	6.6
18	2.4	42	7.3

Source: ASTM C828-80, Low Pressure Air Test of Vitrified Clay Pipe Lines. *Copyright ASTM INTERNATIONAL. Reprinted with permission.*

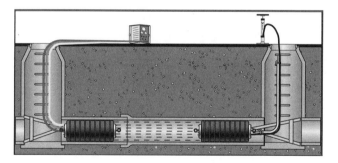

Collection

Courtesy of Cherne Industries.

Low-Pressure Air Testing to Determine the Structural Integrity of Newly Constructed Sewer Lines

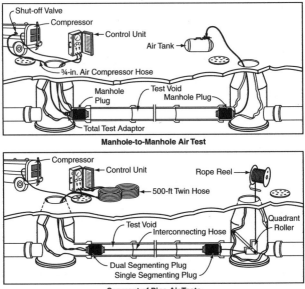

Manhole-to-Manhole Air Test

Segment of Pipe Air Tests

Segment of Pipe Air Tests

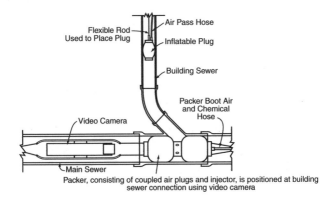

Packer, consisting of coupled air plugs and injector, is positioned at building sewer connection using video camera

Building Sewer Air Testing With Video Camera and Packer (pressure grouting unit)

WATER EXFILTRATION

Test Procedures

1. Plug the downstream end of the sewer line to be tested at the inlet into a manhole and plug the inlet sewer at the upstream manhole.

2. Fill the upstream manhole with water to 4 ft above the invert of the manhole channel, or 4 ft above the average depth of groundwater above the sewer line.

3. Mark the water height in the manhole at the start of the test. Record the time (in hours and minutes).

4. Allow the water to drop 2 ft and record the elapsed time, or wait 2 hours and mark the water height at the end of that time.

5. Measure the diameter of the manhole at the high- and low-water marks. Compute the volume of water (in gallons) that exfiltrated from the sewer line.

6. Compute the rate of water exfiltration (in gallons per minute). Divide the water loss by the time (in minutes) between the start and finish of the test.

7. Obtain gallons per day lost. Multiply the water loss by 1,440 min/day. To determine the rate of water exfiltration (in gallons per day per inch of diameter per mile length of sewer line), divide this amount by the diameter of the pipe (in inches) and the length (in miles). The acceptable rate of water exfiltration from a sewer line is approximately 450 gpd/(in.·mi), or less.

8. Cautiously remove the downstream plug (the release of the test water causes a rapid flow of water in the sewer line). Remove the upstream plug. If the test water is to be reused for the next downstream test section, install another plug at the downstream end of the next test section before releasing water from the bottom plug of the upstream test section.

Collection

Conversion of Infiltration or Exfiltration Rates From Gallons per Inch of Diameter per Mile per Day to Gallons per Hour per 100 ft for Various Sized Sewer Pipes

Diameter of Sewer, *in.*	Filtration Rates in gph/100 ft for the Following Rates, gal/in. diameter/mi/day				
	100	200	300	400	500
8	0.63	1.3	1.9	2.5	3.2
10	0.79	1.6	2.4	3.2	4.0
12	0.95	1.9	2.8	3.8	4.7
15	1.2	2.4	3.5	4.7	5.9
18	1.4	2.8	4.3	5.7	7.1
21	1.7	3.3	5.0	6.6	8.3
24	1.9	3.8	5.7	7.6	9.5
27	2.1	4.3	6.4	8.5	10.7
30	2.4	4.7	7.1	9.5	11.8
36	2.8	5.7	8.5	11.4	14.2
42	3.3	6.6	10.0	13.3	16.6
48	3.8	7.6	11.4	15.2	19.0

Courtesy of Pearson Education, Inc.

PIPE CLEANING AND MAINTENANCE _____

Common Sewer Cleaning Methods

Technology	Uses and Applications
Mechanical	
Rodding	• Uses an engine and a drive unit with continuous rods or sectional rods.
	• As blades rotate, they break up grease deposits, cut roots, and loosen debris.
	• Rodders also help thread the cables used for video inspections and bucket machines.
	• Most effective in lines up to 12 in. (300 mm) in diameter.
Bucket machine	• Cylindrical device, closed on one end with two opposing hinged jaws at the other.
	• Jaws open and scrape off the material and deposit it in the bucket.
	• Partially removes large deposits of silt, sand, gravel, and some types of solid waste.
Hydraulic	
Balling	• A threaded rubber cleaning ball that spins and scrubs the pipe interior as flow increases in the sewer line.
	• Removes deposits of settled inorganic material and grease buildup.
	• Most effective in sewers ranging in size from 5 to 24 in. (13 to 60 cm).
Flushing	• Introduces a heavy flow of water into the line at a manhole.
	• Removes floatables and some sand and grit.
	• Most effective when used in combination with other mechanical operations, such as rodding or bucket machine cleaning.
Jetting	• Directs high velocities of water against pipe walls.
	• Removes debris and grease buildup, clears blockages, and cuts roots within small-diameter pipes.
	• Efficient for routine cleaning of small-diameter, low-flow sewers.
Scooter	• Round, rubber-rimmed, hinged metal shield that is mounted on a steel framework on small wheels. The shield works as a plug to build a head of water.
	• Scours the inner walls of the pipe lines.
	• Effective in removing heavy debris and cleaning grease from line.

Table continued on next page

Collection

Common Sewer Cleaning Methods (continued)

Technology	Uses and Applications
Kites, bags, and poly pigs	• Similar in function to the ball.
	• Rigid rims on bag and kite induce a scouring action.
	• Effective in moving accumulations of decayed debris and grease downstream.
Silt traps	• Collect sediments at convenient locations.
	• Must be emptied on a regular basis as part of the maintenance program.
Grease traps and sand/oil interceptors	• The ultimate solution to grease buildup is to trap and remove it.
	• These devices are required by some uniform building codes and/or sewer-use ordinances. Typically sand/oil interceptors are required for automotive business discharge.
	• Necessary to be thoroughly cleaned to function properly.
	• Cleaning frequency varies from twice a month to once every 6 months, depending on the amount of grease in the discharge.
	• Need to educate restaurant and automobile businesses about the need to maintain these traps.
Chemicals	
(Before using chemicals, review the material safety data sheets [MSDSs] and consult the local authorities on the proper use of chemicals as per local ordinance and the proper disposal of the chemicals used in the operation. If assistance or guidance is needed regarding the application of certain chemicals, contact the USEPA or state water pollution control agency.)	• Used to control roots, grease, odors (H_2S gas), concrete corrosion, rodents, and insects.
	• Root Control—longer lasting effects than power rodder (approximately 2–5 years).
	• H_2S gas—some common chemicals used are chlorine (Cl_2), hydrogen peroxide (H_2O_2), pure oxygen (O_2), air, lime ($Ca(OH_2)$), sodium hydroxide ($NaOH$), and iron salts.
	• Grease and soap problems—some common chemicals used are bioacids, digesters, enzymes, bacteria cultures, catalysts, caustics, hydroxides, and neutralizers.

Limitations of Cleaning Methods

Cleaning Method	Limitation
Balling, jetting, scooter	In general, these methods are only successful when necessary water pressure or head is maintained without flooding basements or houses at low elevations.
	Jetting—The main limitation of this technique is that cautions need to be used in areas with basement fixtures and in steep-grade hill areas.
	Balling—Balling cannot be used effectively in pipes with bad offset joints or protruding service connections because the ball can become distorted.
	Scooter—When cleaning larger lines, the manholes need to be designed to a larger size in order to receive and retrieve the equipment. Otherwise, the scooter needs to be assembled in the manhole. Caution also needs to be used in areas with basement fixtures and in steep-grade hill areas.
Bucket machine	This device has been known to damage sewers. The bucket machine cannot be used when the line is completely plugged because this prevents the cable from being threaded from one manhole to the next. Setup of this equipment is time consuming.
Flushing	This method is not very effective in removing heavy solids. Flushing does not remedy this problem because it only achieves temporary movement of debris from one section to another in the system.
High-velocity cleaner	The efficiency and effectiveness of removing debris by this method decreases as the cross-sectional areas of the pipe increase. Backups into residences have been known to occur when this method has been used by inexperienced operators. Even experienced operators require extra time to clear pipes of roots and grease.
Kite or bag	When using this method, use caution in locations with basement fixtures and steep-grade hill areas.
Rodding	Continuous rods are harder to retrieve and repair if broken, and they are not useful in lines with a diameter of greater than 0.984 ft (300 mm) because the rods have a tendency to coil and bend. This device also does not effectively remove sand or grit but may only loosen the material to be flushed out at a later time.

Collection

Effectiveness of Cleaning Techniques*

Solution to Problem	Emergency Stoppages	Grease	Roots	Sand, Grit, Debris	Odors
Balling		4		4	3
High-velocity cleaning	1	5		4	3
Flushing					2
Sewer scooters		3		3	
Bucket machines, scrapers				2	
Power rodders	4	1	3		
Hand rods	3	1	2		
Chemicals		2	3		3

* On a scale of 1–5, 1 is the least effective solution for a particular problem and 5 is the most effective solution.

Frequency of Maintenance Activities

Activity	Average, % of system/year
Cleaning	29.9
Root removal	2.9
Manhole inspection	19.8
Closed-circuit video inspection	6.8
Smoke testing	7.8

Limitations of Standard Inspection Techniques

Inspection Technique	Limitation
Visual inspection	In smaller sewers, the scope of problems detected is minimal because the only portion of the sewer that can be seen in detail is near the manhole. Therefore, any definitive information on cracks or other structural problems is unlikely. However, this method does provide information needed to make decisions on rehabilitation.
Camera inspection	When performing a camera inspection in a large-diameter sewer, the inspection crew is essentially taking photographs haphazardly, and as a result, the photographs tend to be less comprehensive.
Closed-circuit video	This method requires late night inspection and, as a result, the video operators are vulnerable to lapses in concentration. These inspections are also quite expensive and time consuming.
Lamping inspection	The video camera does not fit into the pipe, and during the inspection, it remains only in the maintenance hole. As a result, only the first 10 ft of the pipe can be viewed or inspected using this method.

Pipe Maintenance

Round Stock Corkscrew: Used for rodding through sewers where conditions are unknown.

Square Stock Corkscrew: Used for removing heavy root growth. Sharpened cutting edge will tear loose roots and remove other rigid obstructions when pulled backward.

Double Corkscrew: A double-pronged tool used for removing miscellaneous obstructions.

Double Sand Corkscrew: The boring action of the corkscrew helps to pull rod through lines impacted with sand, gravel, and similar buildups. This tool must be kept moving since it may settle into builtup material and become stuck.

Auger: Useful for cutting long stringy roots and loosening sedimentary deposits in sewer pipe.

Sewer Rodding Tools and Uses

Collection

Sand Leader: Used to guide rods across the top of built-up materials in the line by the slipping action of the blades.

Root Saw: Used for power sawing of stubborn root masses in the sewer pipe.

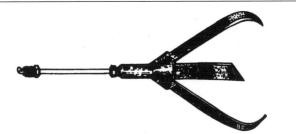

Spring Blade Root Cutter Chuck: This cutter (with appropriately sized blades) is used in preventive maintenance work in sewers. It should be rotated at high speed (power rodding machine) and *pulled* slowly through the line while rotating to effect a thorough scouring of the pipe. NOTE: This tool is not designed to be pushed into a sewer line.

Porcupine: The turn-type porcupine is used in lines up to 12 in. in diameter. Its function is to scour lines of light buildups in conjuction with water flushing of sewer lines.

Sewer Rodding Tools and Uses (continued)

Spearhead Blade: Used in small pipes for removing hard deposits and breaking up hard obstructions such as glass, bottles, cans, and plaster.

Bullet Nose: Designed to be screwed into end of a coupler for least resistance when rodding through heavy roots.

Pickup Tool: Used to snare broken sectional sewer rods.

Assembly Wrench: Used for holding and turning nuts and couplers in assembly rods and tools.

Ratchet Turning Handle: Used with locking pin through pullout tool and coupler to turn rods.

Sewer Rodding Tools and Uses (continued)

Pullout Tool: Used to encircle rod coupler to push rods into or pull rods out of line.

Assembly-Turning Handle: Used for assembling nuts and couplers for turn rods; the spring loader pin engages hole in coupler.

Bar-Turning Handle: Used to secure into hole in coupler for turning, pushing, and pulling rods.

Rod End Swivel: Used for pulling cables and wires through a pipe; designed to be free turning under load at the end coupling.

Sewer Rodding Tools and Uses (continued)

Comparison of Various Sewer Rehabilitation Techniques

	Method	Diameter Range, in. (mm)	Maximum Installation, ft (m)	Liner Material*
In-line expansion	Pipe bursting	4–24 (100–600)	750 (230)	PE, PP, PVC, GRP
Sliplining	Segmental	4–158 (100–4,000)	1,000 (300)	PE, PP, PVC, GRP
	Continuous	4–63 (100–1,600)	1,000 (300)	PE, PP, PE/EPDM, PVC
	Spiral wound	6–100 (150–2,500)	1,000 (300)	PE, PVC, PP, PVDF
Cured-in-place product linings	Inverted-in-place	4–108 (100–2,700)	3,000 (900)	Thermoset Resin/fabric Composite
	Winched-in-place	4–54 (100–1,400)	500 (150)	Thermoset Resin/fabric Composite
	Spray-on linings	3–180 (76–4,500)	500 (150)	Epoxy Resins/cement Mortar

Table continued on next page

Collection

181

Comparison of Various Sewer Rehabilitation Techniques (continued)

Method		Diameter Range, in. (mm)	Maximum Installation, ft (m)	Liner Material*
Modified cross-section methods	Fold and form	4–15 (100–400)	700 (210)	PVC (thermoplastics)
	Deformed/reformed	4–15 (100–400)	2,500 (800)	HDPE (thermoplastics)
	Drawdown	3–24 (62–600)	1,000 (300)	HDPE, MDPE
	Rolldown	3–24 (62–600)	1,000 (300)	HDPE, MDPE
	Thin-walled lining	20–46 (500–1,100)	3,000 (960)	HDPE
Internal point repair	Robotic repair	8–30 (200–760)	NA	Epoxy resins Cement mortar
	Grouting/sealing and spray-on	NA	NA	Chemical grouting
	Link seal	4–24 (100–600)	NA	Special sleeves
	Point cured-in-place product (CIPP)	4–24 (100–600)	50 (15)	Fiberglass/polyester, etc.

Courtesy of the National Utility Contractors Association.

NOTE: Spiral-wound sliplining, robotic repair, and point CIPP can only be used with gravity pipeline. All other methods can be used with both gravity and pressure pipeline.

* EPDM = ethylene-propylene-dienemonomer; GRP = glassfiber-reinforced polyester; HDPE = high-density polyethylene; MDPE = medium-density polyethylene; PE = polyethylene; PP = polypropylene; PVC = polyvinyl chloride; PVDF = polyvinylidene difluoride. N/A = not applicable.

Limitations of Trenchless Sewer Rehabilitation

Method	Limitations
Pipe bursting	• Bypass or diversion of flow required.
	• Insertion pit required.
	• Percussive action can cause significant ground movement. May not be suitable for all materials.
Sliplining	• Insertion pit required.
	• Reduces pipe diameter.
	• Not well suited for small-diameter pipes.
Cured-in-place product (CIPP)	• Bypass or diversion of flow required.
	• Curing can be difficult for long pipe segments.
	• Must allow adequate curing time.
	• Defective installation may be difficult to rectify.
	• Resin may clump together on bottom of pipe.
	• Reduces pipe diameter.
Modified cross section	• Bypass or diversion of flow required.
	• The cross section may shrink or unfold after expansion.
	• Reduces pipe diameter.
	• Infiltration may occur between liner and host pipe unless sealed.
	• Liner may not provide adequate structural support.

Collection

Pumps

Two basic categories of pumps are used in wastewater operations: velocity pumps and positive-displacement pumps. Velocity pumps, which include centrifugal and vertical turbine pumps, are used for most wastewater distribution system applications. Positive-displacement pumps are most commonly used in wastewater treatment plants for chemical metering.

ELECTRICAL MEASUREMENTS

A simple explanation of electrical measurements can be made by comparing the behavior of electricity to the behavior of water.

- Volts (potential) can be compared to the pressure in a water pipe (psi).
- Amperage (current) can be compared to quantity of flow in a pipe (gpm).
- Resistance (ohms) can be likened to the friction loss in a pipe.

FREQUENTLY USED FORMULAS

$$\text{kilowatts} = \frac{\text{disk-watt hours constant} \times \text{revolutions} \times 3{,}600}{\text{seconds} \times 100}$$

1 horsepower = 746 W power

1 horsepower = 0.746 kW power

$$\% \text{ efficiency} = \frac{\text{horsepower output}}{\text{horsepower supplied}} \times 100$$

$$\% \text{ efficiency} = \frac{\text{brake horsepower}}{\text{motor horsepower}} \times 100$$

$$\% \text{ efficiency} = \frac{\text{water horsepower}}{\text{brake horsepower}} \times 100$$

$$\% \text{ efficiency} = \frac{\text{water horsepower}}{\text{motor horsepower}} \times 100$$

power, ft-lb/min = head, ft × flow rate, lb/min)

$$\text{pumping rate} = \frac{\text{gallons}}{\text{minute}}$$

volts = amperes × resistance

watts = volts × amperes

watts = amperes^2 × resistance

$$\text{water horsepower} = \frac{\text{flow rate, gpm} \times \text{total head, ft}}{3{,}960}$$

Single-Phase Alternating Current (AC) Motor

$$\begin{aligned} \text{horsepower} \\ \text{(output)} \end{aligned} = \frac{\text{volts} \times \text{amps} \times \text{efficiency \%} \times \text{power factor}}{746}$$

$$\text{kilowatts} = \frac{\text{volts} \times \text{amps} \times \text{power factor}}{1{,}000}$$

Two-Phase AC Motor

$$\text{kilowatts} = \frac{\text{volts} \times \text{amps} \times \text{power factor}}{1{,}000}$$

Three-Phase AC Motor

$$\begin{aligned} \text{horsepower} \\ \text{(output)} \end{aligned} = \frac{1.73 \times \text{volts} \times \text{amps} \times \text{efficiency \%} \times \text{power factor}}{746}$$

$$\text{kilowatts} = \frac{1.73 \times \text{amps} \times \text{power factor} \times \text{volts}}{1{,}000}$$

Sludge Pumping Head Loss

The Hazen–Williams calculation for head loss is based on a fluid in turbulent flow. As the solids in sludge increase, the fluid becomes increasingly thicker, changing the fluid characteristics and increasing the velocity required for the fluid to become turbulent. Velocities of 5–6 ft/sec are used as an economic balance between pipe size and water head loss. Because sludge lines are rarely sized to be less than 6 in. (150 mm) to prevent clogging and ease cleaning, velocities of less than 2 ft/sec are common. The figure provides a comparison between the flow of water and that of sludge. Water has a shear stress of zero; therefore, at even the smallest amount of energy, water will flow. Sludge will not flow until a threshold amount of pressure or yield stress is applied. Even when it is moving, the amount of energy required to increase sludge velocity is greater than for water and is defined by the coefficient of rigidity.

The Bingham plastic model is a good predictor of sludge head loss in laminar flow. The equation may be written as follows:

$$H/L = \frac{16S_y}{3wD} + \frac{\eta V}{wD^2}$$

Where:

D = diameter of pipe, in ft

S_y = shear stress at the yield point where sludge begins to flow, in lb/ft^2

η = coefficient of ridigity, in lb/ft·sec

H = head loss measured in feet of water height

L = length of pipe, in ft

v = average velocity, in ft/sec

w = weight of water, 64.4 lb/cu ft

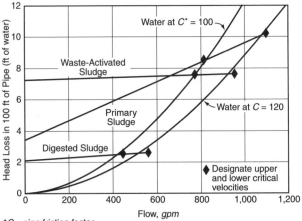

*C = pipe friction factor.

Note: Curves are plotted for waste-activated, digested, and primary sludges at 3.5% water at C = 100 and water at C = 120 with no solids.

Courtesy of Pearson Education, Inc.

Head Loss Versus Flow for 100 ft of 6-in. Pipe

HORSEPOWER AND EFFICIENCY

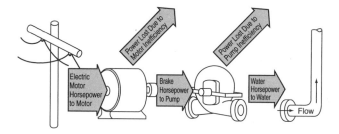

Power Loss Due to Motor and Pump Inefficiency

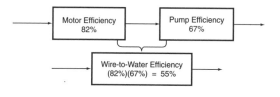

Wire-to-Water Efficiency

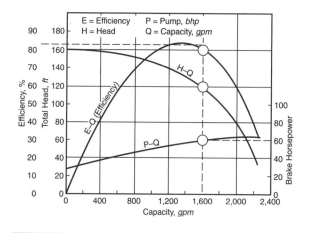

Example Pump Performance Curve

Pumps

Approximate Full Load Current and Fuse Size Required by AC Motors*

hp	115 V			230 V, Single-Phase		
	Amperes	Ordinary Fuse	Time Delay Fuse	Amperes	Ordinary Fuse	Time Delay Fuse
1/6	4.4	15	8	2.2		
1/4	5.8	20	10	2.9		
1/3	7.2	25	12	3.6	16	6
1/2	9.8	30	15	4.9	25	8
3/4	13.8	45	20	6.9	25	12
1	16	50	25	8	25	15
1 1/2	20	60	30	10	30	15
2	24	80	35	12	40	20
3				17	60	25
5				28	90	40
7 1/2				40	125	60
10				50	150	80

* Assumes motors running at usual speeds, with normal torque characteristics.

Three-Phase Induction Motors

hp	220 V			460 V		
	Amperes	Ordinary Fuse	Time Delay Fuse	Amperes	Ordinary Fuse	Time Delay Fuse
1/2	2	15	4	1	15	2
3/4	2.8	15	4	1.4	15	3
1	3.6	15	6	1.8	15	3
1 1/2	5.2	15	8	2.6	15	4
2	6.8	25	10	3.4	15	5
3	9.6	30	15	4.8	15	8
5	15.2	50	25	7.6	25	15
7 1/2	22	75	35	11	35	20
10	28	90	40	14	45	20
15	42	125	60	21	70	30
20	54	175	80	27	90	40
25	68	225	100	34	110	50
30	80	250	125	40	125	60
40	104	350	150	52	175	80
50	130	400	200	65	200	100
60	154	500	250	77	250	125
75	192	600	300	96	300	150
100	248	—	400	124	400	200
125	312	—	450	156	500	250
150	360			180	600	300
200	480			240	—	400

Standard Classification of NEMA Enclosures for Nonhazardous Locations*

Type	Intended Use
1	Intended for indoor use, primarily to provide a degree of protection from persons or equipment contacting the electrical components.
2	Intended for indoor use, to provide some protection against limited amounts of falling water and dirt.
3	Intended for outdoor use, primarily to provide a degree of protection against windblown dust, rain and sleet, and ice on the enclosure.
3R	Intended for outdoor use, primarily to provide a degree of protection against falling rain and sleet; undamaged by the formation of ice on the enclosure.
4	Intended for indoor or outdoor use, primarily to provide a degree of protection against windblown dust and rain, splashing water, and hose-directed water; undamaged by the formation of ice on the enclosure.
4X	Intended for indoor or outdoor use, primarily to provide a degree of protection against corrosion, windblown dust and rain, splashing water, and hose-directed water; undamaged by the formation of ice on the enclosure.
6	Intended for use indoors or outdoors where occasional submersion is encountered.
12	Intended for indoor use, primarily to provide a degree of protection against dust, galling dirt, and dripping noncorrosive liquids.
13	Intended for indoor use, primarily to provide a degree of protection against dust, spraying of water, oil, and noncorrosive coolant.

* These descriptions are in summary form only and are not complete representations of the National Electric Manufacturers Association (NEMA) standards for enclosures.

Pumps

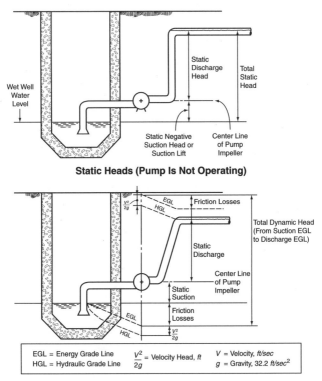

Static Heads (Pump Is Not Operating)

Dynamic Heads (Pump Is Operating)

NOTE: This figure illustrates a pump with a suction lift. Pumps should have a suction head which means the wet well water level should be higher than the pump impeller. This pump will have difficulty starting unless it is a self-priming pump because the water level in the wet well is below the pump. Also, if air gets into the suction line, the only way it can get out is through the pump. Controls may be modified to allow the pump to operate only when a suction head exists if flooding of the service area will not result.

Static and Dynamic Heads

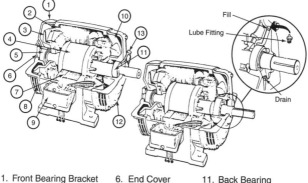

1. Front Bearing Bracket
2. Front Air Deflector
3. Fan
4. Rotor
5. Front Bearing

6. End Cover
7. Stator
8. Screens
9. Conduit Box
10. Back Air Deflector

11. Back Bearing
12. Back Bearing Bracket
13. Oil Lubricuation Cap

Electric Motor Lubrication

New
Smooth surface. May be bright or dull and somewhat discolored due to oxidation or tarnishing.

Used
Surface may be pitted and have discolored areas of black, brown, or may have blue (heat) tint. If half of the thickness (mass) of the silver points is still intact, they are usable. This is the time to order a backup set.

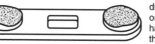

Severe or Long-time Use
Surface badly pitted and eroded with badly feathered and lifting edges. Replace entire contact set.

Pumps

Visual Inspection of Contact Points

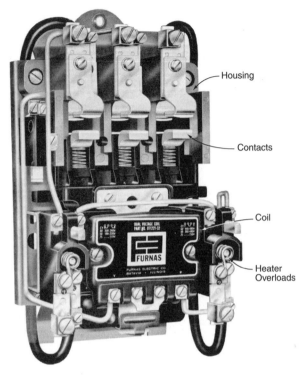

Housing

Contacts

Coil

Heater
Overloads

Photo supplied by Siemens Energy and Automation, Inc.

Three-Phase Magnetic Starter

PUMP VOLAGE

North American Standard System Voltages

Minimum Tolerable	Minimum Favorable	Nominal System	Maximum Favorable	Maximum Tolerable	Type (phase) of System
107	110	120	125	127	1
200	210	240	240	250	3
214/428	220/440	240/480	250/500	254/508	1
244/422	250/434	265/460	227/480	288/500	3
400	420	480	480	500	3
2,100	2,200	2,400	2,450	2,540	3
3,630	3,810	4,160	4,240	4,400	3
6,040	6,320	6,900	7,050	7,300	3
12,100	12,600	13,200	13,800	14,300	3
12,600	13,000	14,400	14,500	15,000	3
30,000		34,500		38,000	3
60,000		69,000		72,500	3
100,000		115,000		121,000	3
120,000		138,000		145,000	3
140,000		161,000		169,000	3

Pumps

North American Standard Nominal Voltages

Nominal System	Generator Rated	Transformer Secondary	Switchgear Rated	Capacitor Rated	Motor Rated	Starter Rated	Ballast Rated
			Single-Phase Systems				
120	120	120	120	—	115	115	118
120/240	120/240	120/240	240	230	230	230	236
208/120	208/120	208/240	240	230	115	115	118
			Three-Phase Systems				
240	240	240	240	230	240	220	236
480/277	480/277	480/277	480	460	460	440	460
480	480/277	480/277	480	480	460	440	460
2,400	2,400/1,388	2,400	2,400	2,400	2,300	2,300	—
4,160	4,160/2,400	4,160/2,400	4,160	4,160	4,000	4,000	—
6,900	6,900/3,980	6,900/3,980	7,200	6,640	6,600	6,600	—
7,200	6,900/3,980	7,200/4,160	13,800	7,200	7,200	7,200	—
12,000	12,500/7,210	12,000/6,920	13,800	12,470	11,000	11,000	—
13,200	13,800/7,970	13,800/7,610	13,800	13,200	13,200	13,200	—
14,400	14,000/8,320	13,800/7,970	14,400	14,400	13,200	13,200	—

Current Ratings for Low-Voltage Switches, in amperes

120/240 V	230 V	240 V	600 V
30	30	30	30
60	60	60	60
100	100	100	100
200	200	200	200
	400	400	400
	600	600	600
	800	800	800
	1,200	1,200	1,200

MAINTENANCE AND TROUBLESHOOTING

Pump and Motor Maintenance Checklist

Refer to the manufacturer's operations and maintenance recommendations for specific guidance. These suggestions are general in nature. The type of equipment that is in operation determines how and when maintenance takes place. Water quality and equipment history play a predominant role in scheduling maintenance. Above all, safety is the main concern when performing any duty on equipment. Electrical, mechanical, and confined-space safety practices must be a part of any preventive maintenance checklist.

Daily (or during routine visits when pump is in operation)

1. Visually observe pump and motor operation.
2. Read the amperage, voltage, flows, run hours, and other information from motor control center.
3. Inspect mechanical seals.
4. Check operating temperature.
5. Check warning indicator lights.
6. Check oil levels.
7. Note any unusual vibration.

Weekly

1. Test per-square-inch levels of the relief valve system; these should be set just above the normal operating pressure of the system.

2. Inspect stuffing box and note the amount of leakage; adjust or lubricate packing gland as necessary. A leakage rate of 20 to 60 drops of seal water per minute is normal for a properly adjusted gland; inadequate or excessive leakage are signs of trouble. Do not overtighten packing gland bolts. Clean drain line if necessary.

3. Check valve lubricant levels.

4. Test the priming system and perform preventive maintenance as necessary.

5. Inspect motor for indications of overload or electrical failure. Check for burnt insulation, melted solder, or discoloration around terminals and wires.

6. Check for and remove any obstructions in or around the impeller, screens, or intake, as appropriate. (Be sure to shut off the pump.)

7. Test transfer valve, if applicable.

Monthly

1. Check bearing temperatures with a thermometer.

2. Clean strainers on system piping including strainers on automatic control valves.

3. Perform dry vacuum test.

4. Check oil level in pump gearbox; add oil as necessary.

5. Inspect gaskets.

6. Check motor ventilation screens and clean or replace as necessary.

7. Check pressure gauge reliability.

8. Check foundation bolts.

9. Clean pump control sensors (may be required weekly, depending on water quality).

10. Check drive flange bolts, if applicable, and tighten as necessary.

Semiannually

1. Perform pump and motor performance test. Check at least three performance test points and plot on the pump's performance curve. Compare this data to the design specifications. Capacity and efficiency determine the degree of pump maintenance necessary.
2. Check pump–motor shaft alignment.
3. Calibrate gauges as necessary.
4. Record vibration levels using vibration level test equipment.
5. Note condition of pump casing, base, and foundation, and of pipe supports and bracing. Correct any deficiencies.
6. Calibrate meters, level sensors, controls, and recording devices as necessary.
7. Inspect and clean check valves, pump control valves, wear rings, and individual drain lines.
8. Inspect intake or screen; replace or clean as necessary.
9. Inspect condition of the impeller, pump shaft, and shaft sleeve; replace as necessary.
10. Lubricate power transfer cylinder, power shift cylinder, shift control valve, and transfer valve mechanism (on two-stage pumps).

Annually

1. Analyze changes in daily data readings.
2. Determine pumping capacity.
3. Determine pumping efficiency.
4. Check all other pumping performance levels, including engine speed and pump pressure.

Pumps

Pump Troubleshooting Guide

Symptom	Probable Cause	Corrective Action
Pump will not start.	Circuit breaker or overload relay tripped, motor cold	Reset breaker or reset manual overload relay.
	Fuses burned out	Check for cause and correct. Replace fuses.
	No power to switch box	Confirm with voltmeter by checking incoming power source. Notify power company.
	Motor is hot and overload relay has tripped.	Allow motor to cool. Check supply voltage. If normal, reset overload relay, start motor, check amperage. If above normal, call electrician.
	Loose or broken wire or short	Tighten wiring terminals. Replace any broken wires. Check for shorts and correct.
	Low line voltage	Check incoming power; use voltmeter. If low, notify power company.
	Defective motor	Meg* out motor. If bad, replace.
	Defective pressure switch	With contact points closed, check for voltage through switch. If no voltage, replace switch; if low voltage, clean contact points; if full voltage, proceed to next item.
	Line to pressure switch is plugged or valve in line has been accidentally shut off.	Open valve if closed. Clean or replace line.

Table continued on next page

Pump Troubleshooting Guide (continued)

Symptom	Probable Cause	Corrective Action
	Pump control valve malfunctioning	Check limit switch for proper travel and contact. Adjust or replace as required.
	Defective time delay relay or pump start timer	Check for voltage through relay or timer; replace as necessary. Check for loose linkage.
	Float switch or transducer malfunctioning	If pump is activated by float switch or pressure transducer on storage tank, check for incoming signal. If no signal, check out switch or transducer with voltmeter. If okay, look for broken cable between storage tank and pump station.
Pump will not shut off.	Defective pressure switch	Points in switch stuck or mechanical linkage broken; replace switch.
	Line to pressure switch is plugged or valve in line has been accidentally shut off.	Open valve if closed. Clean or replace plugged line.
	Cut-off pressure setting too high	Adjust setting.
	Pump control valve malfunctioning	Check limit switch for proper travel and contact. Adjust or replace as required.
	Float switch or transducer malfunctioning	Defective incoming signal; check and replace components as required. Check cable.
	Defective timer in pump stop mode	Check for voltage through pump stop timer; replace if defective.

Table continued on next page

Pumps

Pump Troubleshooting Guide (continued)

Symptom	Probable Cause	Corrective Action
Pump starts too frequently.	Pressure switch cut-in and cut-off settings too close	Adjust settings. Maintain minimum 20 psi (138 kPa or 1.4 kg/cm^2) differential.
	Water-logged tank	Add air to tank. Check air charging system and air release valve. Also check tank and connections for air leaks.
	Leaking foot valve	Check for backflow into well. If excessive or if pump shaft is turning backward, correct problem as soon as possible.
	Time delay relay or pump start/stop timers are malfunctioning.	Check relay or timers for proper operation. Replace defective components.
Fuses blow, circuit breaker or overload relays trip when pump is in operation.	Switch box or control not properly vented, in full sunshine, or in dead air location.	Provide adequate ventilation (may require small fan) and shelter from sun. Paint box or panel with heat-reflective paint, preferably white.
	Overload relay may be tripping because of external heat.	
	Incorrect voltage	Check incoming power source. If not within prescribed limits, notify power company.
	Overload relays tripped	Check motor running amperage. Verify that thermal relay components are correctly sized to operating conditions. Repeated tripping will weaken units. Replace if necessary.

Table continued on next page

Pump Troubleshooting Guide (continued)

Symptom	Probable Cause	Corrective Action
	Motor overloaded and running very hot	Motors are designed to run hot. If the hand can be held on the motor for 10 seconds without extreme discomfort, the temperature is not damaging. Motor current should not exceed nameplate rating. Fifteen percent overload reduces motor life by 50%.
Pump will not deliver normal amount of water.	Pump breaking suction	Check water level to be certain water is above pump bowls when operating. If not, lower bowls.
	Pump impellers improperly adjusted	Check adjustment and lower impellers (qualified personnel only).
	Rotation incorrect	Check rotation.
	Impellers worn	If well pumps sand, impellers could be excessively worn, reducing amount of water pump can deliver. Evaluate and recondition pump bowls if required.
	Pump control valve malfunctioning	Check limit switch for proper travel and contact. Adjust or replace as required.
	Impeller or bowls partially plugged	Wash down pump by forcing water back through discharge pipe. Evaluate sand production from well.
	Drawdown more than anticipated	Check pumping water level. Reduce production from pump or lower bowls.

Table continued on next page

Pumps

203

Pump Troubleshooting Guide (continued)

Symptom	Probable Cause	Corrective Action
	Pump motor speed too slow	Check speed and compare with performance curves. Also check lift and discharge pressure for power requirements.
Pump takes too much power.	Impellers not properly adjusted	Refer to manufacturer's bulletin for adjustment of open or closed impellers.
	Well is pumping sand.	Check water being pumped for presence of sand. Restrict discharge until water is clear. Care should be taken not to shut down pump if it is pumping very much sand.
	Crooked well, pump shaft binding	Reshim between pump base and pump head to center shaft in motor quill. Never shim between pump head and motor.
	Worm bearings or bent shaft	Check and replace as necessary.
Excessive operating noise.	Motor bearings worn	Replace as necessary.
	Bent line shaft or head shaft	Check and replace.
	Line shaft bearings not receiving oil	Make sure there is oil in the oil reservoir and that the oiler solenoid is opening. Check sight gauge drip rate. Adjust drip feed oiler for 5 drops/min plus 1 drop/min for each 40 ft (12 m) of column.

Source: Office of Water Programs, California State University, Sacramento Foundation, in Small Water System Operation and Maintenance. For additional information, visit <www.owp.csus.edu> or call 916-278-6142.

* Meg is a procedure used for checking the insulation resistance on motors, feeders, buss bar systems, grounds, and branch circuit wiring.

Pump Troubleshooting

Indication	Suggestions
No leakage at startup	Back off gland to encourage generous leakage. If negative suction, install lantern and connect to discharge.
Excessive leakage at startup	Check for correct packing size. Were rings installed correctly in accordance with instructions? Check pump for run-out.
Packing rings flattened out on inside diameter under the rod or shaft	Check bearings. Packing probably supporting the shaft weight.
Packing rings flattened out above rod or shaft, or on either side of same	Check alignment of shaft. Worn bearings may be causing whip or run-out.
Distinct bulge on side of ring	Probably too wide a gap on adjacent ring—rings cut too short
Sides of rings shiny or worn	Rings may be too loose and are rotating with the shaft.
Rings extruding past gland follower	Too much clearance between outside diameter of shaft and inside diameter of follower. Apply bushing. May also be excessive gland pressure.
Top or gland end rings in poor condition; bottom rings okay	Set improperly installed
Rings disappear in set.	Packing entering the system. Install bottom bushing.
Packing is torn.	Check sleeve for burrs. Are there any abrasives?
Rings are burned; faces dried and charred.	Check for proper size packing. Is packing the correct selection with respect to heat limitations and/or peripheral speed? Check lubrication. Are there any abrasives?
Packing hardens.	See suggestions for "burned rings" above. Are there congealing liquids involved?
Packing softens.	Check for proper style selection, also for lubrication.
Excessive loss of lubricant.	Excessive gland pressure. Check for packing selection, temperature, whip in shaft.
Unexplained leakage.	Shaft sleeve may be leaking. Replace sleeve seal under sleeve.
Packing freezes to shaft after shutdown.	Liquid salting out or congealing in packing set. Provide lubrication to packing before shutdown.

Pumps

Troubleshooting Guide for Electric Motors and Solutions

Symptoms	Cause	Result*	Remedy
1. Motor does not start. (Switch is on and not defective.)	a. Incorrectly connected	a. Burnout	a. Connect correctly per diagram on motor.
	b. Incorrect power supply	b. Burnout	b. Use only with correctly rated power supply.
	c. Fuse out, loose, or open connection	c. Burnout	c. Correct open-circuit condition.
	d. Rotating parts of motor may be mechanically jammed .	d. Burnout	d. Check and correct: • Bent shaft • Broken housing • Damaged bearing • Foreign material in motor
	e. Driven machine may be jammed.	e. Burnout	e. Correct jammed condition.
	f. No power supply	f. None	f. Check for voltage at motor and work back to power supply.
	g. Internal circuitry open	g. Burnout	g. Correct open-circuit condition.
2. Motor starts but does not come up to speed.	a. Same as 1a, b, c	a. Burnout	a. Same as 1a, b, c
	b. Overload	b. Burnout	b. Reduce load to bring current to rated limit. Use proper fuses and overload protection.
	c. One or more phases out on a three-phase motor	c. Burnout	c. Look for open circuits.

Table continued on next page

Troubleshooting Guide for Electric Motors and Solutions (continued)

Symptoms	Cause	Result*	Remedy
3. Motor noisy electrically	Same as 1a, b, c	Burnout	Same as 1a, b, c
4. Motor runs hot (exceeds rating).	a. Same as 1a, b, c	a. Burnout	a. Same as 1a, b, c
	b. Overload	b. Burnout	b. Reduce load.
	c. Impaired ventilation	c. Burnout	c. Remove obstruction.
	d. Frequent start or stop	d. Burnout	d. Reduce number of starts or reversals. Secure proper motor for this duty.
	e. Misalignment between rotor and stator laminations	e. Burnout	e. Realign.
5. Noisy (mechanically)	a. Misalignment of coupling or sprocket	a. Bearing failure, broken shaft, stator burnout due to motor drag	a. Correct misalignment.
	b. Mechanical unbalance of rotating parts	b. Same as 5a	b. Find unbalanced part, then balance.
	c. Lack of or improper lubricant	c. Bearing failure	c. Use correct lubricant; replace parts as necessary.
	d. Foreign material in lubricant	d. Same as 5c	d. Clean out and replace bearings.
	e. Overload	e. Same as 5c	e. Remove overload condition. Replace damaged parts.

Table continued on next page

Pumps

Troubleshooting Guide for Electric Motors and Solutions (continued)

Symptoms	Cause	Result*	Remedy
	f. Shock loading	f. Same as 5c	f. Correct causes and replace damaged parts.
	g. Mounting acts as amplifier of normal noise.	g. Annoying	g. Isolate motor from base.
	h. Rotor dragging due to worn bearings, shaft, or bracket	h. Burnout	h. Replace bearings, shaft, or bracket as needed.
6. Bearing failure	a. Same as 5a, b, c, d, e	a. Burnout, damaged shaft, damaged housing	a. Replace bearings and follow 5a, b, c, d, e.
	b. Entry of water or foreign material into bearing housing	b. Same as 6a	b. Replace bearings and seals and shield against entry of foreign material (water, dust, etc.). Use proper motor.

* Many of these conditions should trip protective devices rather than burn out motors.

Troubleshooting Guide for Electric Motors

Symptoms	Caused By	Appearance
Shorted motor winding	Moisture, chemicals, foreign material in motor, damaged winding	Black or burned coil with remainder of winding good
All windings completely burned	Overload	Burned equally all around winding
	Stalled	Burned equally all around winding
	Impaired ventilation	Burned equally all around winding
	Frequent reversal or starting	Burned equally all around winding
	Incorrect power	Burned equally all around winding
Single-phase condition	Open circuit in one line. The most common causes are loose connection, one fuse out, loose contact in switch.	If 1,800-rpm motor—four equally burned groups at 90° intervals
		If 1,200-rpm motor—six equally burned groups at 60° intervals
		If 3,600-rpm motor—two equally burned groups at 180°
		Note: If Y connected, each burned group will consist of two adjacent phase groups. If delta connected, each burned group will consist of one-phase group.
Other	Improper connection	Irregularly burned groups or spot burns
	Ground	

Note: Many burnouts occur within a short period of time after motor is started up. This does not necessarily indicate that the motor was defective, but usually is due to one or more of the above-mentioned causes. The most common of these are improper connections, open circuits in one line, incorrect power supply, or overload.

Pumps

Split Packing Box

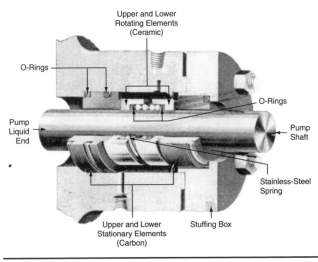

Mechanical Seal

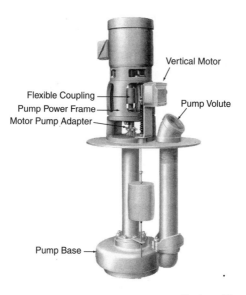

Courtesy of Aurora Pump.

Flexibly Coupled, Vertically Mounted Centrifugal Pump

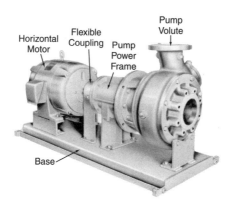

Courtesy of Aurora Pump.

Flexibly Coupled, Horizontally Mounted Centrifugal Pump

Pumps

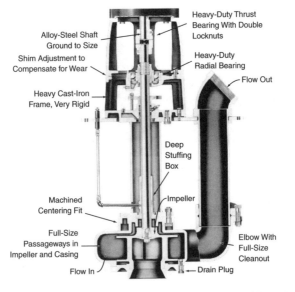

Courtesy of Aurora Pump.

Vertical Ball-bearing–Type Wastewater Pump

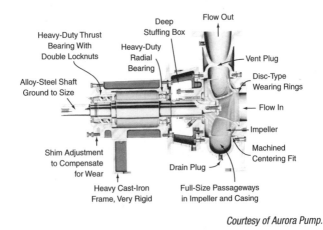

Courtesy of Aurora Pump.

Horizontal Nonclog Wastewater Pump With Open Impeller

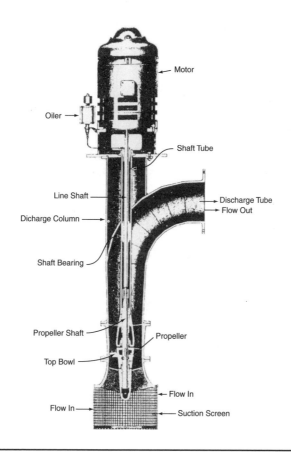

Propeller Pump

1. Lifting Handle
2. Junction Chamber With Watertight Cable Entries
3. Antifriction Bearings
4. Shaft
5. Stator With Temperature Sensing Thermistors
6. Rotor
7. Brg. Temperature Thermistor
8. Stator Housing Leakage Sensor
9. Oil Chamber
10. Shaft Seal
11. Pump Housing (Volute)
12. Nonclog Impeller
13. Cooling Jacket
14. Sliding Bracket
15. Automatic Discharge Connection

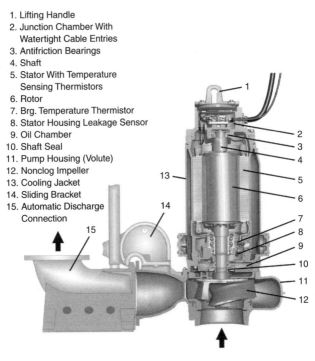

Courtesy of ITT Flygt Corporation.

Submersible Wastewater Pump

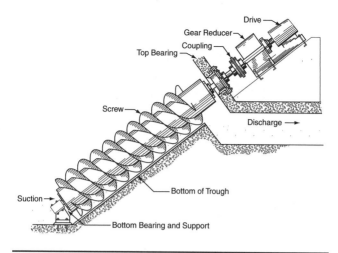

Incline Screw Pump

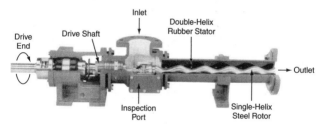

Courtesy of Moyno, Inc. a unit of Robbins and Myers, Inc.
"Moyno" is a registered tradmark of Moyno, Inc.

Cutaway View of a Progressive Cavity Pump Used for Pumping Sludge

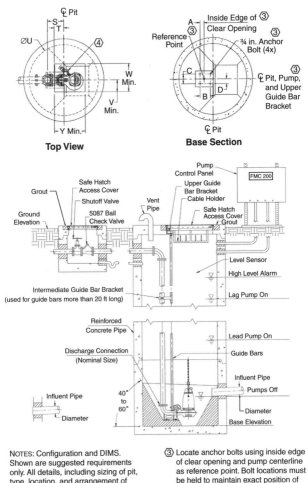

Top View

Base Section

NOTES: Configuration and DIMS. Shown are suggested requirements only. All details, including sizing of pit, type, location, and arrangement of valves and piping, etc., are to be specified by the consulting engineer and are subject to his or her approval.

③ Locate anchor bolts using inside edge of clear opening and pump centerline as reference point. Bolt locations must be held to maintain exact position of pump to clear opening.

④ Mix—flush valve.

Courtesy of ITT Flygt Corporation.

Submersible Pump in Wet Well

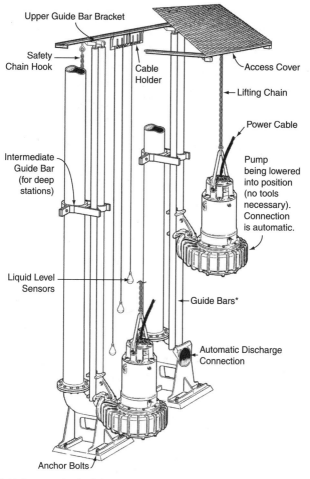

Upper Guide Bar Bracket

Safety Chain Hook

Cable Holder

Access Cover

Lifting Chain

Power Cable

Pump being lowered into position (no tools necessary). Connection is automatic.

Intermediate Guide Bar (for deep stations)

Liquid Level Sensors

Guide Bars*

Automatic Discharge Connection

Anchor Bolts

*Guide bars are standard pipe.

Courtesy of ITT Flygt Corporation.

Duplex Submersible Pumping Station

Pumps

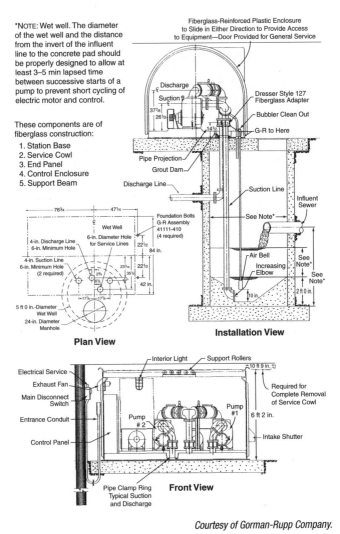

*NOTE: Wet well. The diameter of the wet well and the distance from the invert of the influent line to the concrete pad should be properly designed to allow at least 3–5 min lapsed time between successive starts of a pump to prevent short cycling of electric motor and control.

These components are of fiberglass construction:

1. Station Base
2. Service Cowl
3. End Panel
4. Control Enclosure
5. Support Beam

Fiberglass-Reinforced Plastic Enclosure to Slide in Either Direction to Provide Access to Equipment—Door Provided for General Service

Discharge
Suction
Dresser Style 127 Fiberglass Adapter
Bubbler Clean Out
G-R to Here
Pipe Projection
Grout Dam
Discharge Line
Suction Line
Influent Sewer
See Note*
Air Bell
Increasing Elbow
See Note*
See Note*

Plan View

76¾   47¼
Wet Well
4-in. Discharge Line 6-in. Minimum Hole
6-in. Diameter Hole for Service Lines
4-in. Suction Line 6-in. Minimum Hole (2 required)
Foundation Bolts G-R Assembly 41111-410 (4 required)
5 ft 0 in.-Diameter Wet Well
24-in. Diameter Manhole
84 in.
42 in.

Installation View

Front View

Interior Light   Support Rollers
Electrical Service
Exhaust Fan
Main Disconnect Switch
Entrance Conduit
Control Panel
Pump #2
Pump #1
10 ft 9 in.
Required for Complete Removal of Service Cowl
6 ft 2 in.
Intake Shutter
Pipe Clamp Ring Typical Suction and Discharge

Courtesy of Gorman-Rupp Company.

Above-Ground Duplex Pump Station With a Fiberglass-Reinforced Plastic Enclosure

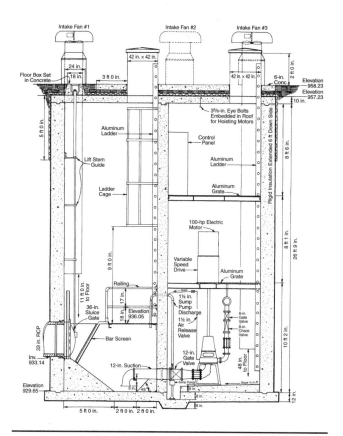

Section of Cast-in-Place Lift Station

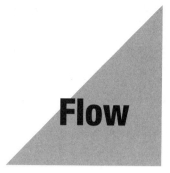

Flow

The movement of water and wastewater is dynamic with many variables for monitoring and measuring flow. Maintaining the proper flow is critical to wastewater operations. Wastewater uses specific devices unique to the industry.

Summary of Pressure Requirements

Requirement	Value		Location
	psi	*(kPa)*	
Minimum pressure	35	(241)	All points within distribution system
	20	(140)	All ground level points
Desired maximum	100	(690)	All points within distribution system
Fire flow minimum	20	(140)	All points within distribution system
Ideal range	50–75	(345–417)	Residences
	35–60	(241–414)	All points within distribution system

KEY CONVERSIONS FOR FLOWS

Conversion of US customary flow units can be easily made using the block diagram below. When moving from a smaller to a larger block, multiply by the factor shown on the connecting line. When moving from a larger to a smaller block, divide.

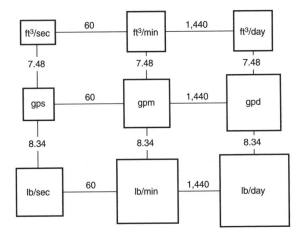

$$\text{flow, gpm} = \text{flow, cfs} \times 448.8 \text{ gpm/cfs}$$

$$\text{flow, cfs} = \frac{\text{flow, gpm}}{448.8 \text{ gpm/cfs}}$$

pipe diameter, in. $= \dfrac{\text{area, ft}^2}{0.785} \times 12$ in./ft

actual leakage, gpd/mi./in. $= \dfrac{\text{leak rate, gpd}}{\text{length, mi.} \times \text{diameter, in.}}$

NOTE: minimum flushing velocity: 2.5 fps

maximum pipe velocity: 5.0 fps

key conversions: 1.55 cfs/mgd; 448.8 gpm/cfs

KEY FORMULAS FOR FLOWS AND METERS

Velocity

flow, cfs $=$ area, ft $\times$ velocity, fps

$$\dfrac{\text{gpm}}{448.8 \text{ gpm/cfs}} = 0.785 \times \text{diameter, ft}^2 \times \dfrac{\text{distance, ft}}{\text{time, sec}}$$

velocity, fps $= \dfrac{\text{flow, cfs}}{\text{area, ft}^2}$

area, ft^2 $= \dfrac{\text{flow, cfs}}{\text{velocity, fps}}$

Head Loss Resulting From Friction

Darcy-Weisbach Formula

$$h_L = f(L/D)(V^2/2g)$$

Where (in any consistent set of units):

h_L = head loss

f = friction factor, dimensionless

L = length of pipe

D = diameter of the pipe

V = average velocity

g = gravity constant

Flow

Hazen–Williams Formula

$$h_f = k_1 \frac{L Q^{1.85}}{C^{1.85} D^{4.87}}$$

Where:

h_f = head loss, in ft

k_1 = 4.72, in units of seconds$^{1.85}$ per feet$^{0.68}$

L = pipe length, in ft

Q = flow rate, in cfs

C = Hazen–Williams roughness coefficient

D = pipe diameter, in ft

The value of C ranges from 60 for corrugated steel to 150 for clean, new asbestos–cement pipe.

Manning Formula

$$v = \frac{1.486}{n} R^{\frac{2}{3}} S^{\frac{1}{2}}$$

Where:

v = flow velocity, in fps

n = Manning coefficient of channel roughness

R = hydraulic radius, in ft

S = channel slope (for uniform flow) or the energy slope (for nonuniform flow), dimensionless

The energy slope is calculated as $-dH/dx$, where H is the total energy, which is expressed as

$$H = z + y + \frac{v^2}{2g}$$

Where (in any consistent set of units):

x = elevation head

y = water depth

v = velocity

g = gravitational constant

x = distance between any two points

Approximate Flow Through Venturi Tube

$$Q = 19.05\, d_1^2 \sqrt{H} \sqrt{\frac{1}{1 - \left(\dfrac{d_1}{d_2}\right)^4}}$$

for any Venturi tube.

$$Q = 19.17\, d_1^2 \sqrt{h}$$

for a Venturi tube in which $d_1 = 1/3\, d_2$.

Where:

Q = flow, in gpm

d_1 = diameter of Venturi throat, in in.

H = difference in head between upstream end and throat, in ft

d_2 = diameter of main pipe, in in.

These formulas are suitable for any liquid with viscosities similar to water. The values given here are for water. A value of 32.174 ft/sec^2 was used for the acceleration of gravity and a value of 7.48 gal/ft^3 was used in computing the constants.

Flow

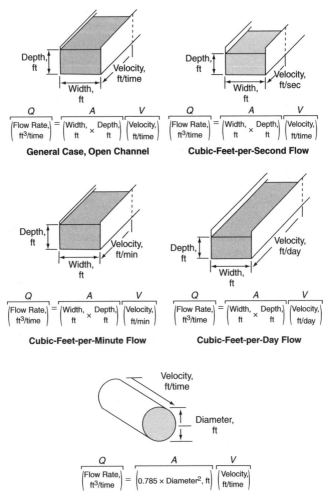

General Case, Open Channel

$$\underset{\substack{\left(\text{Flow Rate,}\atop \text{ft}^3/\text{time}\right)}}{Q} = \underset{\substack{\left(\text{Width,}\atop \text{ft} \times \text{Depth,}\atop \text{ft}\right)}}{A} \underset{\substack{\left(\text{Velocity,}\atop \text{ft/time}\right)}}{V}$$

Cubic-Feet-per-Second Flow

$$\underset{\substack{\left(\text{Flow Rate,}\atop \text{ft}^3/\text{time}\right)}}{Q} = \underset{\substack{\left(\text{Width,}\atop \text{ft} \times \text{Depth,}\atop \text{ft}\right)}}{A} \underset{\substack{\left(\text{Velocity,}\atop \text{ft/time}\right)}}{V}$$

Cubic-Feet-per-Minute Flow

$$\underset{\substack{\left(\text{Flow Rate,}\atop \text{ft}^3/\text{time}\right)}}{Q} = \underset{\substack{\left(\text{Width,}\atop \text{ft} \times \text{Depth,}\atop \text{ft}\right)}}{A} \underset{\substack{\left(\text{Velocity,}\atop \text{ft/min}\right)}}{V}$$

Cubic-Feet-per-Day Flow

$$\underset{\substack{\left(\text{Flow Rate,}\atop \text{ft}^3/\text{time}\right)}}{Q} = \underset{\substack{\left(\text{Width,}\atop \text{ft} \times \text{Depth,}\atop \text{ft}\right)}}{A} \underset{\substack{\left(\text{Velocity,}\atop \text{ft/day}\right)}}{V}$$

General Case, Circular Pipe Flowing Full

$$\underset{\substack{\left(\text{Flow Rate,}\atop \text{ft}^3/\text{time}\right)}}{Q} = \underset{\left(0.785 \times \text{Diameter}^2, \text{ft}\right)}{A} \underset{\substack{\left(\text{Velocity,}\atop \text{ft/time}\right)}}{V}$$

The $Q = AV$ Equation As It Pertains to Flow in an Open Channel

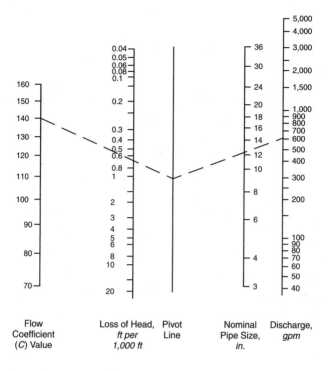

Flow Coefficient (C) Value	Loss of Head, ft per 1,000 ft	Pivot Line	Nominal Pipe Size, in.	Discharge, gpm

Draw a line between two known values and extend it so that it touches the pivot line. Draw a line between the point on the pivot line and the other known value. Read the unknown value where the second line intersects the graph.

Flow of Water in Ductile-Iron Pipe

Flow

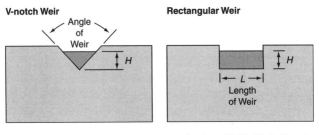

V-notch Weir

Rectangular Weir

Courtesy of Public Works Magazine.

Types of Weirs

The two most commonly used weir types are the V-notch and rectangular, illustrated in the figure above. To read a flow rate graph or table pertaining to a weir, you must know two measurements: (1) the height H of the water above the weir crest; and (2) the *angle* of the weir (V-notch weir) *or* the *length* of the crest (rectangular weir).

$$\text{weir overflow rate} = \frac{\text{flow, gpd}}{\text{weir length, ft}}$$

Example

A nomograph for 60° and 90° V-notch weirs is given in the figure on page 229. Using this nomograph, determine (a) the flow rate in gallons per minute if the height of water above the 60° V-notch weir crest is 12 in.; and (b) the gallons-per-day flow rate over a 90° V-notch weir when the height of the water is 12 in. over the crest.

(a) The scales used on this graph are logarithmic. This information is important because it determines how interpolation should be performed when the indicated flow falls between two known values.

First, draw a horizontal line from 12 on the *height* scale on the left to 12 on the *height* scale on the right. Then on the scale for a 60° V-notch, read the flow rate

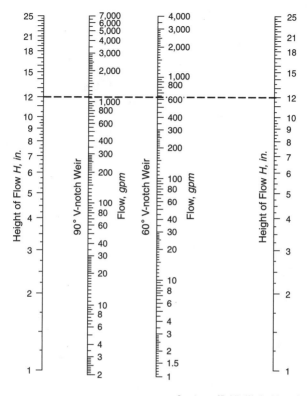

Flow Rate Nomograph for 60° and 90° V-notch Weirs

indicated by a 12-in. head. The flow rate falls between 600 and 700 gpm at approximately 650 gpm.

(b) On the scale for a 90° V-notch, the indicated flow rate is between 1,000 and 2,000 gpm. More precisely, it falls between 1,100 and 1,200 gpm at a reading of about 1,150 gpm. Convert the gallons-per-minute rate to gallons per day:

$$(1,150 \text{ gpm})(1,440 \text{ min/day}) = 1,656,000 \text{ gpd}$$

Discharge From a V-Notch Weir With End Contractions*

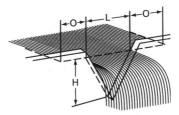

Head (H)		Discharge Over Weir, *gpm*				
		Weir Angle, *degrees*				
in.	10th of foot	22.5	30	45	60	90
1	.083	0.4	0.5	0.8	1.2	2.0
1¼	.104	0.8	1.0	1.6	2.2	3.9
1½	.125	1.2	1.7	2.6	3.5	6.1
1¾	.146	1.8	2.4	3.8	5.2	9.1
2	.167	2.6	3.4	5.3	7.3	12.7
2¼	.188	3.4	4.6	7.1	9.8	17.1
2½	.208	4.4	5.9	9.1	12.7	22.0
2¾	.229	5.6	7.5	11.6	16.1	27.9
3	.250	7.0	9.4	14.4	20.1	34.8
3¼	.271	8.7	11.4	17.9	24.9	43.1
3½	.292	10.3	13.8	21.3	29.6	51.3
3¾	.313	12.3	15.4	25.3	35.2	61.0
4	.333	14.4	19.2	29.6	41.1	71.2
4¼	.354	16.7	22.3	34.5	47.8	83.0
4½	.375	19.3	25.8	39.8	55.3	95.8
4¾	.396	22.1	29.5	45.6	63.3	109.9
5	.417	25.2	33.6	51.8	71.9	124.8
5¼	.437	28.3	37.8	58.4	81.1	140.6
5½	.458	31.9	42.5	65.6	91.1	158.0
5¾	.479	35.6	47.4	73.3	101.7	176.4
6	.500	39.7	53.0	81.8	113.6	196.9

*The distance (O) on either side of the weir must be at least ¾ L.

Example

The table on page 232 pertains to the discharge of 45° V-notch weirs. Use the table to determine (a) flow rate in cubic feet per second when the head above the crest is 0.75 ft; (b) the gallons-per-day flow rate when the head is 1.5 ft.

(a) In the table, part of the head (0.7) is given on the vertical scale, and the remainder (0.05) is given on the horizontal scale (0.7 + 0.05 = 0.75). The cubic-feet-per-second flow rate indicated by a head of 0.75 ft is 0.504 ft³/sec.

(b) A head of 1.5 ft is read as 1.5 on the vertical scale and 0.00 on the horizontal scale (1.5 + 0.00 = 1.50). The million-gallons-per-day flow rate indicated by this head is 1.84 mgd. This is equal to a flow rate of 1,840,000 gpd. (The *mgd* column was read in this problem because it is easier to convert to gallons per day from million gallons per day than from cubic feet per second.)

Flow

Discharge of 45° V-notch Weirs*

Head, ft	.00 ft³/sec	.00 mgd	.01 ft³/sec	.01 mgd	.02 ft³/sec	.02 mgd	.03 ft³/sec	.03 mgd	.04 ft³/sec	.04 mgd	.05 ft³/sec	.05 mgd	.06 ft³/sec	.06 mgd	.07 ft³/sec	.07 mgd	.08 ft³/sec	.08 mgd	.09 ft³/sec	.09 mgd
0.1	.003	.002	.004	.003	.005	.003	.006	.004	.008	.005	.009	.006	.011	.007	.012	.008	.014	.009	.016	.011
0.2	.019	.012	.021	.014	.024	.015	.026	.017	.029	.019	.032	.021	.036	.023	.039	.025	.043	.028	.047	.030
0.3	.051	.033	.055	.036	.060	.039	.065	.042	.070	.045	.075	.048	.081	.052	.086	.056	.092	.060	.098	.064
0.4	.105	.068	.111	.072	.118	.077	.126	.081	.133	.086	.141	.091	.149	.096	.157	.101	.165	.107	.174	.112
0.5	.183	.118	.192	.124	.202	.130	.212	.137	.222	.143	.232	.150	.243	.157	.254	.164	.265	.171	.277	.179
0.6	.289	.187	.301	.194	.313	.203	.326	.211	.339	.219	.353	.228	.366	.237	.380	.246	.395	.255	.410	.265
0.7	.425	.274	.440	.284	.455	.294	.471	.305	.488	.315	.504	.326	.521	.337	.539	.348	.556	.360	.574	.371
0.8	.593	.383	.611	.395	.630	.407	.650	.420	.670	.433	.690	.446	.710	.459	.731	.472	.752	.486	.774	.500
0.9	.796	.514	.818	.529	.841	.543	.864	.558	.887	.573	.911	.589	.935	.604	.960	.620	.985	.636	1.01	.653
1.0	1.04	.669	1.06	.686	1.09	.703	1.11	.721	1.14	.738	1.17	.756	1.20	.774	1.23	.793	1.25	.811	1.28	.830

Table continued on next page

Discharge of 45° V-notch Weirs* (continued)

Head, ft	.00		.01		.02		.03		.04		.05		.06		.07		.08		.09	
	ft³/sec	mgd	ft³/sec	mgd	ft³/sec	mgd	ft³/sec	mgd	ft³/sec	mgd	ft³/sec	mgd	ft³/sec	mgd	ft³/sec	mgd	ft³/sec	mgd	ft³/sec	mgd
1.1	1.31	.849	1.34	.869	1.37	.888	1.41	.908	1.44	.929	1.47	.949	1.50	.970	1.53	.991	1.57	1.01	1.60	1.03
1.2	1.63	1.06	1.67	1.08	1.70	1.10	1.74	1.12	1.77	1.15	1.81	1.17	1.84	1.19	1.88	1.22	1.92	1.24	1.96	1.26
1.3	1.99	1.29	2.03	1.31	2.07	1.34	2.11	1.36	2.15	1.39	2.19	1.42	2.23	1.44	2.27	1.47	2.32	1.50	2.36	1.52
1.4	2.40	1.55	2.44	1.58	2.49	1.61	2.53	1.64	2.58	1.66	2.62	1.69	2.67	1.72	2.71	1.75	2.76	1.78	2.81	1.81
1.5	2.85	1.84	2.90	1.87	2.95	1.91	3.00	1.94	3.05	1.97	3.10	2.00	3.15	2.03	3.20	2.07	3.25	2.10	3.30	2.13
1.6	3.35	2.17	3.41	2.20	3.46	2.23	3.51	2.27	3.57	2.30	3.62	2.34	3.68	2.38	3.73	2.41	3.79	2.45	3.84	2.48
1.7	3.90	2.52	3.96	2.56	4.02	2.60	4.08	2.63	4.13	2.67	4.19	2.71	4.25	2.75	4.32	2.79	4.38	2.83	4.44	2.87
1.8	4.50	2.91	4.56	2.95	4.63	2.99	4.69	3.03	4.75	3.07	4.82	3.11	4.89	3.16	4.95	3.20	5.02	3.24	5.08	3.29
1.9	5.15	3.33	5.22	3.37	5.29	3.42	5.36	3.46	5.43	3.51	5.50	3.55	5.57	3.60	5.64	3.64	5.71	3.69	5.78	3.74
2.0	5.86	3.79	5.93	3.83	6.00	3.88	6.08	3.93	6.15	3.98	6.23	4.03	6.31	4.08	6.38	4.13	6.46	4.18	6.54	4.23

Adapted from Leupold and Stevens, Inc., P.O. Box 688, Beaverton, Oregon 97005, from Stevens Water Resources Data Book.

* ft³/sec = 1.035 H^(5/2); mgd = ft³/sec × 0.646.

Flow

233

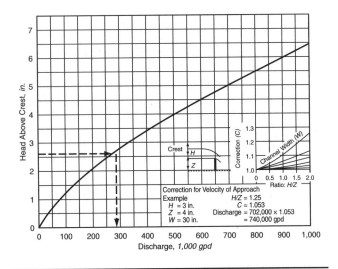

Discharge Curve, 15-in. Contracted Rectangular Weir

Example

A discharge curve for a 15-in. contracted rectangular weir is given in the figure above. Using this curve, determine the gallons-per-day flow rate if the head above the weir crest is 2.75 in.

The scales for the discharge curve are arithmetic, so it is much easier to determine the indicated flow rate when the arrow falls between two known values. In this case, a head of 3.75 ft indicates a flow rate between 250,000 and 300,000 gpd, or about 285,000 gpd.

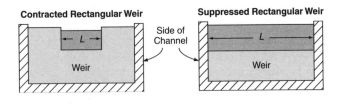

Two Types of Rectangular Weirs

Discharge From a Rectangular Weir With End Contractions*

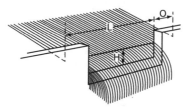

Head (H)		Discharge Over Weir, *gpm*			
		Length (L) of Weir, *ft*			
in.	*10th of foot*	1	3	5	Additional gpm for Each Foot Over 5 ft
1	.083	35.4	107.5	179.8	36.05
1¼	.104	49.5	150.4	250.4	50.4
1½	.125	64.9	197	329.5	66.2
1¾	.146	81	240	415	83.5
2	.167	98.5	302	506	102
2¼	.188	117	361	605	122
2½	.208	136.2	422	706	143
2¾	.229	157	485	815	165
3	.250	177.8	552	926	187
3¼	.271	199.8	624	1,047	211
3½	.292	222	695	1,167	236
3¾	.313	245	769	1,292	261
4	.333	269	846	1,424	288
4¼	.354	293.6	925	1,559	316
4½	.375	318	1,006	1,696	345
4¾	.936	344	1,091	1,835	374
5	.417	370	1,175	1,985	405
5¼	.437	395.5	1,262	2,130	434
5½	.458	421.6	1,352	2,282	465
5¾	.479	449	1,442	2,440	495
6	.500	476.5	1,535	2,600	528

*The distance (0) on either side of the weir must be at least 3 H.

Flow

Example

A nomograph for rectangular weirs (contracted and suppressed) is shown on page 237. Using this nomograph, determine (a) the flow in gallons per minute over a suppressed rectangular weir if the length of the weir is 3 ft and the height of the water over the weir is 4 in.; (b) the flow in gallons per minute over a contracted rectangular weir for the same weir length and head as in (a).

To use the nomograph, you must know the difference between a contracted rectangular weir (one *with* end contractions) and a suppressed rectangular weir (one *without* end contractions). A contracted rectangular weir comes in somewhat from the side of the channel before the crest cutout begins. On a suppressed rectangular weir, however, the crest cutout stretches from one side of the channel to the other.

(a) To determine the flow over the suppressed weir, draw a line from L = 3 ft on the left-hand scale through H = 4 in. (right side of the middle scale). A flow rate of 850 gpm is indicated where the line crosses the right-hand scale. This is the flow over the suppressed rectangular weir.

(b) To determine the flow rate over a contracted rectangular weir using the nomograph, first determine the flow rate over a suppressed weir given the weir length and head, as in (a). Then subtract the flow indicated on the middle scale.

In this example, the flow rate over a 3-ft-long suppressed weir with a head of 4 in. is 850 gpm. To determine the flow rate over a contracted weir 3 ft long with a head of 4 in., a correction factor must be subtracted from the 850 gpm. As indicated by the middle scale, the correction factor is 20 gpm.

$$
\begin{array}{ll}
850 \text{ gpm} & \text{suppressed rectangular weir} \\
\underline{-\ 20 \text{ gpm}} & \\
830 \text{ gpm} & \text{contracted rectangular weir}
\end{array}
$$

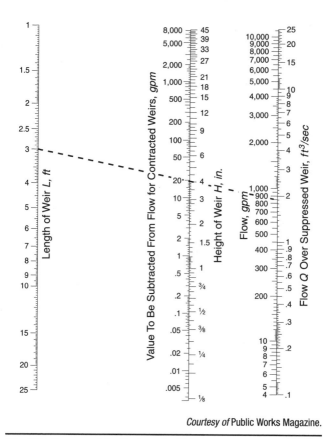

Courtesy of Public Works Magazine.

Flow Rate Nomograph for Rectangular Weirs

Example

Use the table on pages 238–240 to determine the flow rate (in million gallons per day) over a contracted rectangular weir if the length of the weir crest is 3 ft and the head is 0.58 ft.

Enter the table under the *head* column at 0.58; move right until you come under the 3 heading for *length of weir crest*. The indicated flow rate is 2.739 mgd.

Flow Through Contracted Rectangular Weirs

			Length of Weir Crest, ft										
	1		1½		2		3		4		5	
Head, ft	ft³/sec	mgd	ft³/sec	mgd	ft³/sec	mgd	ft³/sec	mgd	ft³/sec	mgd	ft³/sec	mgd
.36	.667	.431	1.026	.663	1.386	.895	2.105	1.360	2.824	1.824	3.543	2.289
.37	.695	.448	1.070	.690	1.445	.932	2.195	1.416	2.945	1.900	3.695	2.384
.38	.721	.465	1.111	.717	1.501	.969	2.281	1.473	3.061	1.976	3.841	2.480
.39	.748	.483	1.153	.745	1.559	1.006	2.370	1.530	3.181	2.054	3.992	2.577
.40	.775	.500	1.196	.772	1.617	1.044	2.459	1.588	3.301	2.132	4.143	2.676
.41	.802	.518	1.239	.800	1.676	1.083	2.550	1.647	3.424	2.216	4.298	2.776
.42	.830	.536	1.283	.829	1.736	1.121	2.642	1.707	3.548	2.292	4.454	2.877
.43	.858	.554	1.327	.857	1.797	1.160	2.736	1.767	3.675	2.373	4.614	2.979
.44	.886	.572	1.372	.886	1.858	1.200	2.830	1.827	3.802	2.455	4.774	3.082
.45	.915	.591	1.417	.915	1.920	1.240	2.925	1.889	3.930	2.538	4.935	3.187
.46	.943	.609	1.462	.945	1.982	1.280	3.021	1.951	4.060	2.621	5.099	3.292
.47	.972	.628	1.508	.974	2.045	1.320	3.118	2.013	4.191	2.706	5.264	3.399
.48	1.001	.646	1.554	1.004	2.108	1.361	3.215	2.076	4.322	2.791	5.429	3.506
.49	1.030	.665	1.601	1.034	2.172	1.403	3.314	2.140	4.456	2.878	5.598	3.615
.50	1.059	.684	1.647	1.064	2.236	1.444	3.413	2.204	4.590	2.965	5.767	3.725

Table continued on next page

Flow Through Contracted Rectangular Weirs (continued)

Head, ft	Length of Weir Crest, ft														
	1		1½		2		3		4		5	
	ft³/sec	mgd	ft³/sec	mgd	ft³/sec	mgd	ft³/sec	mgd	ft³/sec	mgd	ft³/sec	mgd
.51	1.089	.703	1.695	1.095	2.302	1.486	3.515	2.269	4.728	3.052	5.941	3.835
.52	1.119	.722	1.743	1.126	2.368	1.529	3.617	2.335	4.866	3.141	6.115	3.947
.53	1.149	.742	1.791	1.156	2.434	1.571	3.719	2.401	5.004	3.230	6.289	4.060
.54	1.178	.761	1.838	1.188	2.499	1.614	3.820	2.467	5.141	3.321	6.462	4.174
.55	1.209	.781	1.888	1.219	2.567	1.658	3.925	2.534	5.283	3.411	6.641	4.288
.56	1.240	.800	1.938	1.251	2.636	1.701	4.032	2.602	5.428	3.503	6.824	4.404
.57	1.270	.820	1.986	1.282	2.703	1.745	4.136	2.670	5.569	3.595	7.002	4.520
.58	1.300	.840	2.035	1.314	2.771	1.790	4.242	2.739	5.713	3.689	7.184	4.638
.59	1.331	.859	2.085	1.347	2.840	1.830	4.349	2.808	5.858	3.783	7.367	4.757
.60	1.362	.879	2.136	1.380	2.910	1.879	4.458	2.879	6.006	3.877	7.554	4.876
.61	1.393	.899	2.186	1.412	2.980	1.924	4.567	2.948	6.154	3.972	7.741	4.997
.62	1.424	.920	2.237	1.444	3.050	1.970	4.676	3.019	6.302	4.068	7.928	5.118
.63	1.455	.940	2.287	1.477	3.120	2.015	4.785	3.090	6.450	4.165	8.115	5.240
.64	1.487	.960	2.339	1.510	3.192	2.061	4.897	3.162	6.602	4.262	8.307	5.363
.65	1.518	.980	2.390	1.544	3.263	2.107	5.008	3.234	6.753	4.360	8.498	5.487

Table continued on next page

Flow

239

Flow Through Contracted Rectangular Weirs (continued)

Head, ft	Length of Weir Crest, ft									
	1		1½		2		3		4		5	
	ft³/sec	mgd	ft³/sec	mgd	ft³/sec	mgd	ft³/sec	mgd	ft³/sec	mgd	ft³/sec	mgd
.66	1.550	1.001	2.443	1.577	3.336	2.153	5.122	3.306	6.908	4.459	8.694	5.612
.67	1.581	1.021	2.494	1.611	3.407	2.200	5.233	3.379	7.059	4.558	8.885	5.738
.68	1.613	1.042	2.546	1.644	3.480	2.247	5.347	3.453	7.214	4.658	9.081	5.864
.69	1.646	1.062	2.600	1.680	3.555	2.295	5.464	3.527	7.373	4.759	9.282	5.991
.70	1.677	1.083	2.652	1.713	3.627	2.342	5.577	3.601	7.527	4.860	9.477	6.120
.71	1.709	1.104	2.705	1.747	3.701	2.390	5.693	3.676	7.685	4.962	9.677	6.249
.72	1.741	1.124	2.758	1.781	3.775	2.438	5.809	3.751	7.843	5.065	9.877	6.379
.73	1.774	1.145	2.812	1.816	3.851	2.486	5.928	3.827	8.005	5.168	10.08	6.510

Adapted from Leupold and Stevens, Inc., P.O. Box 688, Beaverton, Oregon 97005, from Stevens Water Resources Data Book.

Parshall Flumes

The Parshall flume is another type of flow-metering device designed to measure flows in open channels. Overhead and side views of a flume are shown below.

To use most nomographs pertaining to flow in a Parshall flume, you must know the width of the throat section of the flume (W), the upstream depth of water (depth H_a in the tube of one stilling well), and the depth of water at the foot of the throat section (depth H_b in the tube of the other stilling well).

At normal flow rates, a nomograph may be used directly to determine the flow through the Parshall flume. At higher flows, however, the throat begins to become submerged. When this occurs, a correction to the nomograph reading is required. For a flume that has a 1-in. throat width, this correction factor is not

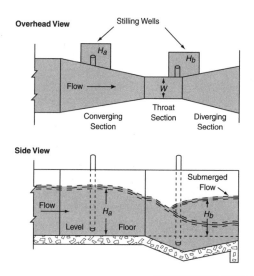

Courtesy of Public Works Magazine.

Parshall Flume

required until the percent of submergence $[(H_b/hr_a) \times 100]$ is greater than 50 percent. For a flume that has a 1-ft throat width, the correction is not required until the percent of submergence exceeds 70 percent. The following examples use the Parshall flume nomograph and corrections graph shown on page 243.

Example. Suppose that your treatment plant has a Parshall flume with a 4-ft throat width. If the upstream depth of the water (depth H_a) is 1 ft and the depth of water in the throat section (depth H_b) is 0.65 ft, what is the gallons-per-minute flow rate through the flume?

First, determine the approximate flow rate using the flume throat width and the upstream (H_a) depth reading. Then use the correction graph on page 243 if required.

Line A is drawn on the nomograph from 4 ft on the *throat width* (left) scale through 1 ft on the H_a *depth* scale (right side of the middle scale); the flow rate indicated where the line crosses the right scale is about 7,200 gpm.

To determine whether a correction to this flow rate is required, first calculate the percent submergence of the throat:

$$
\begin{aligned}
\text{percent submergence} &= \frac{H_b}{H_a} \times 100 \\[2mm]
&= \frac{0.65 \text{ ft}}{1 \text{ ft}} \times 100 \\[2mm]
&= 0.65 \times 100 \\[2mm]
&= 65\%
\end{aligned}
$$

The correction line corresponding to a 4-ft throat width does not even begin until the percent submergence reaches about 78 percent. Therefore, in this example, no correction to the 7,200-gpm flow rate is required.

Example. A flume has a throat width of 3 ft. If the upstream depth of the water (depth H_a) is 9 in. and the depth of water in the throat section (depth H_b) is 7.5 in., what is the gallons-per-day flow rate through the flume?

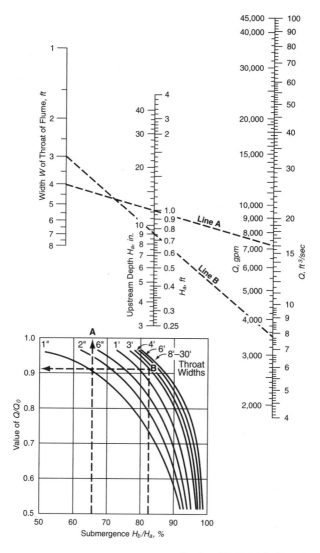

Courtesy of Public Works Magazine.

Parshall Flume Flow Rate Nomograph and Corrections Graph

First determine the approximate flow rate in the flume using the flume throat width and the upstream (H_a) depth reading. Then use the correction graph if needed.

Line B is drawn on the nomograph from 3 ft on the *throat width* (left) scale through 9 in. on the H_a *depth* scale (left side of the middle scale); the flow rate indicated where the line crosses the right scale is 3,400 gpm. Convert this flow rate to gallons per day:

$$(3,400 \text{ gpm})(1,440 \text{ min/day}) = 4,896,000 \text{ gpd}$$

To determine whether a correction to this flow rate is needed, first calculate the percent submergence of the throat:

$$\text{percent submergence} = \frac{H_b}{H_a} \times 100$$

$$= \frac{7.5 \text{ ft}}{9 \text{ in.}} \times 100$$

$$= 0.83 \times 100$$

$$= 83\%$$

Looking at the correction line corresponding to a 3-ft throat width, you can see that a correction is needed once the submergence exceeds about 77 percent. Because the submergence calculated is 83 percent, a correction is required. Draw a vertical line up from the indicated 83 percent until it crosses the 3-ft correction line (point B on the correction graph). Then, move directly to the left until you reach the Q/Q_0 scale. Calculate the corrected flow through the flume by multiplying the flow shown on the nomograph by the Q/Q_0 value.

In this case, the Q/Q_0 value is about 0.92. Multiply this by the flow rate obtained from the nomograph:

$$(4,896,000 \text{ gpd})(0.92) = 4,504,320 \text{ gpd corrected flow}$$

Palmer-Bowlus Flumes

Top View

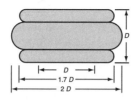

Elevation View

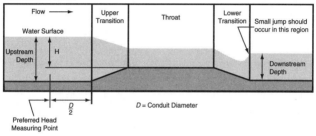

Courtesy of Teledyne Isco, Inc.

Palmer–Bowlus Flumes Are Designed to Be Installed in Existing Channels With Minimal Effort

Table of Flow Data

Size	4 in.	6 in.	8 in.	10 in.	12 in.	15 in.
Maximum mgd	0.419	0.887	1.687	1.896	3.988	5.772
Maximum gpm	291.2	615.8	1,171.5	1,316.5	2,769.1	4,008.2
Maximum cfs	0.649	1.372	2.610	2.933	6.170	8.931

Flow

Venturi Meters

Venturi meters are designed to measure flow in closed-conduit or pressure pipeline systems. As shown in the figure, flow measurement in this type of meter is based on differences in pressure between the upstream section (diameter D_1) and throat section (diameter D_2) of the meter. One type of nomograph available for use with venturi meters is shown in the graph on page 247. The graph applies to meters for which the diameter at D_1 divided by the diameter at D_2 equal to 0.5. It may be used for meters having D_1/D_2 values as high as 0.75, but the results will be about as high as 1 percent. For high flow rates, the nomograph may be used with no correction factors necessary. For low flow rates, however, the correction factors graph shown must be used to find a C' value. The value of Q obtained from the nomograph is then multiplied by the C' value to obtain the final answer.

Example. Based on the nomograph on page 247, what is the flow rate in cubic feet per second in a venturi meter with a 1-in. throat if the difference in head between D_1 and D_2 is 4 ft?

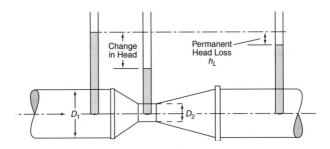

Courtesy of Public Works Magazine.

Venturi Meter

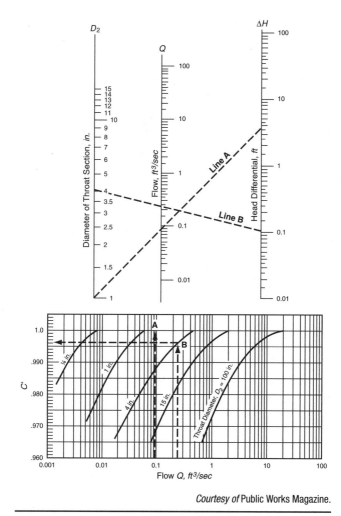

Courtesy of Public Works Magazine.

Venturi Meter Flow Rate Nomograph and Corrections Graph

First, determine the cubic-feet-per-second flow rate using the nomograph, then use a correction factor if needed.

Line A is drawn from 1 in. on the *throat diameter* (left) scale to 4 ft on the *head differential* (right) scale. The flow rate indicated where the line crosses the middle scale is about 0.09 ft³/sec.

To determine if a flow correction is required, first locate 0.09 on the *flow* (bottom) scale of the correction graph. Then move upward until you cross the 1-in. throat diameter line (point A). The flow rate corresponds with a C' value of 1.0. Therefore, the flow obtained from the nomograph requires no correction because

$$\text{corrected flow} = (Q \text{ from nomograph})(C' \text{ from correction graph})$$

$$= (0.09 \text{ ft}^3/\text{sec})(1.0)$$

$$= 0.09 \text{ ft}^3/\text{sec}$$

The corrected flow is the same as the flow obtained from the nomograph.

Examine the correction graph closely, and notice that any flow *at or above* 0.06 ft³/sec for a 1-in. throat diameter would need no correction, because the C' values at these points equal 1.0. Any flow less than 0.06 ft³/sec would require correction.

Example. The venturi meter at your treatment plant has a throat diameter of 4 in. If the pressure differential between D_1 and D_2 is 0.1 ft, what is the flow rate in cubic feet per second?

First determine the cubic-feet-per-second flow rate using the nomograph. Then use a correction factor if necessary.

On the nomograph on page 247, line B is drawn from 4 in. on the *throat diameter* (left) scale to 0.1 ft on the *head differential* (right) scale; the flow rate indicated on the middle scale is about 0.22 ft³/sec.

Now, to determine if a correction to this flow rate is required, locate 0.22 ft³/sec on the *flow* scale of the correction graph and move upward until you intersect the 4-in. throat diameter line at point B. The flow corresponds to a C' value of 0.996. Therefore, the corrected flow rate is

$$\underset{\text{flow}}{\text{corrected}} = (Q \text{ from nomograph})(C' \text{ from correction graph})$$

$$= (0.22 \text{ ft}^3/\text{sec})(0.996)$$

$$= 0.219 \text{ ft}^3/\text{sec}$$

The corrected flow is essentially the same as the flow indicated on the nomograph (0.219 ft³/sec compared with 0.22 ft³/sec).

Orifice Meters

An orifice meter is another type of head differential meter designed for use in closed conduits. As shown below, there is a pressure tap on each side of the orifice plate, and flow rate calculations are based on the difference in pressure between the upstream (high-pressure) tap (point 1) and the downstream (low-pressure) tap (point 2). The nomograph on page 250 can be used to estimate the flow in orifice meters.

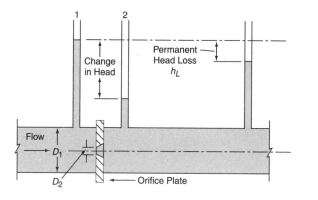

Courtesy of Public Works Magazine.

Orifice Meter

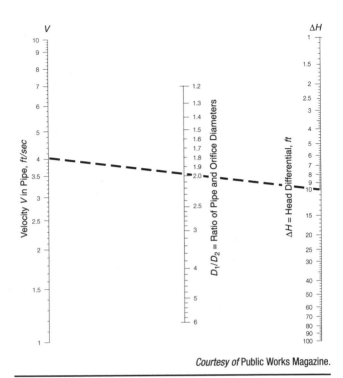

Courtesy of Public Works Magazine.

Flow Rate Nomograph for Orifice Meter

Example. If the head differential between pressure taps 1 and 2 is 10.2 ft, the diameter of the pipe is 6 in., and the diameter of the orifice is 3 in., what is the flow rate in the pipeline in cubic feet per second?

First calculate the ratio of the pipe diameter to the orifice diameter and locate the result on the right side of the middle scale of the nomograph on page 250.

$$\frac{D_1}{D_2} = \frac{6 \text{ in.}}{3 \text{ in.}}$$

$$= 2.0$$

Now draw a line from 10.2 on the right-hand scale, through 2.0 on the right side of the middle scale, and across to the velocity scale. The indicated velocity is 4 ft/sec.

The diameter of the pipeline is known (so that area can be calculated), and the velocity of the flow is known, so use the $Q = AV$ equation to determine the flow rate in the pipe:

$$Q = AV$$

$$= (0.785)(0.5 \text{ ft})(0.5 \text{ ft})(4 \text{ ft/sec})$$

$$= 0.79 \text{ ft}^3/\text{sec}$$

Flow

Flow Equivalents

Cubic Feet per Second

ft³/sec	to	gpm	to	gal/24 hr	to	m³/hr
0.2		90		129,263		20.39
0.4		180		258,526		40.78
0.6		269		387,789		61.17
0.8		359		517,052		81.56
1.0		449		646,315		102.0
1.2		539		775,578		122.3
1.4		628		904,841		142.7
1.6		718		1,034,104		163.1
1.8		808		1,163,367		183.5
2.0		898		1,292,630		203.9
2.2		987		1,421,893		224.3
2.4		1,077		1,551,156		244.7
2.6		1,167		1,680,420		265.1
2.8		1,257		1,809,683		285.5
3.0		1,346		1,938,946		305.9
3.2		1,436		2,068,209		326.2
3.4		1,526		2,197,472		346.6
3.6		1,616		2,326,735		367.0

Gallons per 24 Hours

gal/24 hr	to	gpm	to	ft³/sec	to	m³/hr
100,000		69		0.15		15.77
125,000		87		0.19		19.71
200,000		139		0.31		31.54
400,000		278		0.62		63.08
500,000		347		0.77		78.85
600,000		417		0.93		94.62
700,000		486		1.08		110.4
800,000		556		1.24		126.2
900,000		625		1.39		141.9
1,000,000		694		1.55		157.7
2,000,000		1,389		3.09		315.4
3,000,000		2,083		4.64		473.1
4,000,000		2,778		6.19		630.8
5,000,000		3,472		7.74		788.5
6,000,000		4,167		9.28		946.2
7,000,000		4,861		10.83		1,104
8,000,000		5,556		12.38		1,262
9,000,000		6,250		13.92		1,419

Table continued on next page

Flow Equivalents (continued)

| | Cubic Feet per Second | | | | | | | | Gallons per 24 Hours | | | | | |
ft³/sec	to	gpm	to	gal/24 hr	to	m³/hr		gal/24 hr	to	gpm	to	ft³/sec	to	m³/hr
3.8		1,705		2,455,998		387.4		10,000,000		6,944		15.47		1,577
4.0		1,795		2,585,261		407.8		12,000,000		8,333		18.56		1,892
4.2		1,885		2,714,524		428.2		12,500,000		8,680		19.34		1,971
4.4		1,975		2,843,787		448.6		14,000,000		9,722		21.65		2,208
4.6		2,068		2,973,050		469.0		15,000,000		10,417		23.20		2,366
4.8		2,154		3,102,313		489.4		16,000,000		11,111		24.75		2,523
5.0		2,244		3,231,576		509.8		18,000,000		12,500		26.85		2,839
10.0		4,488		6,463,152		1,020		20,000,000		13,889		30.94		3,154
20.0		8,987		12,926,304		2,039		25,000,000		17,361		38.68		3,943
30.0		13,464		19,389,456		3,059		30,000,000		20,833		46.41		4,731
40.0		17,952		25,852,261		4,078		40,000,000		27,778		61.88		6,308
50.0		22,440		32,315,760		5,098		50,000,000		34,722		77.35		7,885
60.0		26,928		38,778,912		6,117		60,000,000		41,667		92.82		9,462
70.0		31,416		45,242,064		7,137		70,000,000		48,611		108.29		11,039
75.0		33,660		48,473,640		7,646		75,000,000		52,083		116.04		11,828
80.0		35,904		51,705,216		8,156		80,000,000		55,556		123.76		12,616
90.0		40,392		58,168,368		9,176		90,000,000		62,500		139.23		14,193
100.0		44,880		64,631,520		10,195		100,000,000		69,444		154.72		15,770

Table continued on next page

Flow

Flow Equivalents (continued)

Cubic Feet per Second							Gallons per 24 Hours						
ft³/sec	to	gpm	to	gal/24 hr	to	m³/hr	gal/24 hr	to	gpm	to	ft³/sec	to	m³/hr
101.0		45,329		65,277,835		10,297	125,000,000		86,805		193.40		19,713
102.0		45,778		65,924,150		10,399	150,000,000		104,167		232.08		23,665
103.0		46,226		66,570,466		10,501	175,000,000		121,528		270.76		27,598
104.0		46,675		67,216,781		10,603	200,000,000		138,889		309.44		31,540
105.0		47,124		67,863,096		10,705	225,000,000		156,250		348.12		35,483
106.0		47,572		68,509,411		10,807	250,000,000		173,611		386.80		39,425
107.0		48,022		69,155,726		10,909	300,000,000		208,333		464.16		47,310
108.0		48,470		69,802,042		11,011	400,000,000		277,778		618.88		63,080
109.0		48,919		70,448,357		11,113	500,000,000		347,220		773.60		78,850
110.0		49,368		71,094,672		11,215	600,000,000		416,664		928.32		94,620
120.0		53,856		77,557,824		12,234	700,000,000		486,108		1,083.04		110,390
125.0		56,100		80,798,400		12,744	750,000,000		520,328		1,160.40		118,275
130.0		58,344		84,020,976		13,254	800,000,000		555,552		1,237.76		126,160
140.0		62,832		90,484,128		14,273	900,000,000		624,996		1,392.48		141,930
150.0		67,320		96,947,230		15,293	1,000,000,000		694,440		1,547.20		157,700

NOTE: gpm and gal/24 hr given to nearest whole number. The value 7.48 gal = 1 ft³ is used in calculating the values in this table.

254

Flow Data—Nozzles, Theoretical Discharge of Nozzles in US Gallons per Minute (1/16–13/8-in. diameters)

Head		Velocity of Discharge,	Diameter of Nozzle, in.												
psi	ft	ft/sec	1/16	1/8	3/16	1/4	3/8	1/2	5/8	3/4	7/8	1	1 1/8	1 1/4	1 3/8
10	23.1	38.6	0.37	1.48	3.32	5.91	13.3	23.6	36.9	53.1	72.4	94.5	120	148	179
15	34.6	47.25	0.45	1.81	4.06	7.24	16.3	28.9	45.2	65.0	88.5	116.0	147	181	219
20	46.2	54.55	0.52	2.09	4.69	8.35	18.8	33.4	52.2	75.1	102.0	134.0	169	209	253
25	57.7	61.0	0.58	2.34	5.25	9.34	21.0	37.3	58.3	84.0	114.0	149.0	189	234	283
30	69.3	66.85	0.64	2.56	5.75	10.2	23.0	40.9	63.9	92.0	125.0	164.0	207	256	309
35	80.8	72.2	0.69	2.77	6.21	11.1	24.8	44.2	69.0	99.5	135.0	177.0	224	277	334
40	92.4	77.2	0.74	2.96	6.64	11.8	26.6	47.3	73.8	106.0	145.0	188.0	239	296	357
45	103.9	81.8	0.78	3.13	7.03	12.5	28.2	50.1	78.2	113.0	153.0	200.0	253	313	379
50	115.5	86.25	0.83	3.30	7.41	13.2	29.7	52.8	82.5	119.0	162.0	211.0	267	330	399
55	127.0	90.5	0.87	3.46	7.77	13.8	31.1	55.3	86.4	125.0	169.0	221.0	280	346	418
60	138.6	94.5	0.90	3.62	8.12	14.5	32.5	57.8	90.4	130.0	177.0	231.0	293	362	438
65	150.1	98.3	0.94	3.77	8.45	15.1	33.8	60.2	94.0	136.0	184.0	241.0	305	376	455
70	161.7	102.1	0.98	3.91	8.78	15.7	35.2	62.5	97.7	141.0	191.0	250.0	317	391	473
75	173.2	105.7	1.01	4.05	9.08	16.2	36.4	64.7	101.0	146.0	198.0	259.0	327	404	489
80	184.8	109.1	1.05	4.18	9.39	16.7	37.6	66.8	104.0	150.0	205.0	267.0	338	418	505
85	196.3	112.5	1.08	4.31	9.67	17.3	38.8	68.9	108.0	155.0	211.0	276.0	349	431	521

Table continued on next page

Flow

255

Flow Data—Nozzles, Theoretical Discharge of Nozzles in US Gallons per Minute (1/16–13/8-in. diameters) (continued)

Head		Velocity of Discharge,	Diameter of Nozzle, in.												
psi	ft	ft/sec	1/16	1/8	3/16	1/4	3/8	1/2	5/8	3/4	7/8	1	11/8	11/4	13/8
90	207.9	115.8	1.11	4.43	9.95	17.7	39.9	70.8	111.0	160.0	217.0	284.0	359	443	536
95	219.4	119.0	1.14	4.56	10.2	18.2	41.0	72.8	114.0	164.0	223.0	292.0	369	456	551
100	230.9	122.0	1.17	4.67	10.5	18.8	42.1	74.7	117.0	168.0	229.0	299.0	378	467	565
105	242.4	125.0	1.20	4.79	10.8	19.2	43.1	76.5	120.0	172.0	234.0	306.0	388	479	579
110	254.0	128.0	1.23	4.90	11.0	19.6	44.1	78.4	122.0	176.0	240.0	314.0	397	490	593
115	265.5	130.9	1.25	5.01	11.2	20.0	45.1	80.1	125.0	180.0	245.0	320.0	406	501	606
120	277.1	133.7	1.28	5.12	11.5	20.5	46.0	81.8	128.0	184.0	251.0	327.0	414	512	619
130	300.2	139.1	1.33	5.33	12.0	21.3	48.0	85.2	133.0	192.0	261.0	341.0	432	533	645
140	323.3	144.3	1.38	5.53	12.4	22.1	49.8	88.4	138.0	199.0	271.0	354.0	448	553	668
150	346.4	149.5	1.43	5.72	12.9	22.9	51.6	91.5	143.0	206.0	280.0	366.0	463	572	692
175	404.1	161.4	1.55	6.18	13.9	24.7	55.6	98.8	154.0	222.0	302.0	395.0	500	618	747
200	461.9	172.6	1.65	6.61	14.8	26.4	59.5	106.0	165.0	238.0	323.0	423.0	535	660	799
250	577.4	193.0	1.85	7.39	16.6	29.6	66.5	118.0	185.0	266.0	362.0	473.0	598	739	894
300	692.8	211.2	2.02	8.08	18.2	32.4	72.8	129.0	202.0	291.0	396.0	517.0	655	808	977

NOTE: The actual quantity discharged by a nozzle will be less than shown in this table. A well-tapered smooth nozzle may be assumed to give 97% to 99% of the values in the table.

* Where there is both an upstream and downstream pressure, the head is a differential head.

Flow Data—Nozzles, Theoretical Discharge of Nozzles in US Gallons per Minute (1½–6-in. diameters)

| Head* | | Velocity of Discharge, ft/sec | Diameter of Nozzle, in. | | | | | | | | | | | | |
psi	ft		1½	1¾	2	2¼	2½	2¾	3	3½	4	4½	5	5½	6
10	23.1	38.6	213	289	378	479	591	714	851	1,158	1,510	1,915	2,365	2,855	3,405
15	34.6	47.25	260	354	463	585	723	874	1,041	1,418	1,850	2,345	2,890	3,490	4,165
20	46.2	54.55	301	409	535	676	835	1,009	1,203	1,638	2,135	2,710	3,340	4,040	4,810
25	57.7	61.0	336	458	598	756	934	1,128	1,345	1,830	2,385	3,025	3,730	4,510	5,380
30	69.3	66.85	368	501	655	828	1,023	1,236	1,473	2,005	2,615	3,315	4,090	4,940	5,895
35	80.8	72.2	398	541	708	895	1,106	1,335	1,591	2,168	2,825	3,580	4,415	5,340	6,370
40	92.4	77.2	425	578	756	957	1,182	1,428	1,701	2,315	3,020	3,830	4,725	5,610	6,810
45	103.9	81.8	451	613	801	1,015	1,252	1,512	1,802	2,455	3,200	4,055	5,000	6,050	7,120
50	115.5	86.25	475	647	845	1,070	1,320	1,595	1,900	2,590	3,375	4,275	5,280	6,380	7,600
55	127.0	90.4	498	678	886	1,121	1,385	1,671	1,991	2,710	3,540	4,480	5,530	6,690	7,970
60	138.6	94.5	521	708	926	1,172	1,447	1,748	2,085	2,835	3,700	4,685	5,790	6,980	8,330
65	150.1	98.3	542	737	964	1,220	1,506	1,819	2,165	2,950	3,850	4,875	6,020	7,270	8,670
70	161.7	102.1	563	765	1,001	1,267	1,565	1,888	2,250	3,065	4,000	5,060	6,250	7,560	9,000
75	173.2	105.7	582	792	1,037	1,310	1,619	1,955	2,330	3,170	4,135	5,240	6,475	7,820	9,320
80	184.8	109.1	602	818	1,070	1,354	1,672	2,020	2,405	3,280	4,270	5,410	6,690	8,080	9,630
85	196.3	112.5	620	844	1,103	1,395	1,723	2,080	2,480	3,375	4,400	5,575	6,890	8,320	9,920

Table continued on next page

Flow

Flow Data—Nozzles, Theoretical Discharge of Nozzles in US Gallons per Minute (1¹/₂–6-in. diameters) (continued)

| Head* | | Velocity of Discharge, ft/sec | Diameter of Nozzle, in. | | | | | | | | | | | | |
psi	ft		1¹/₂	1³/₄	2	2¹/₄	2¹/₂	2³/₄	3	3¹/₂	4	4¹/₂	5	5¹/₂	6
90	207.9	115.8	638	868	1,136	1,436	1,773	2,140	2,550	3,475	4,530	5,740	7,090	8,560	10,210
95	219.4	119.0	656	892	1,168	1,476	1,824	2,200	2,625	3,570	4,655	5,900	7,290	8,800	10,500
100	230.9	122.0	672	915	1,196	1,512	1,870	2,255	2,690	3,660	4,775	6,050	7,470	9,030	10,770
105	242.4	125.0	689	937	1,226	1,550	1,916	2,312	2,755	3,750	4,890	6,200	7,650	9,260	11,020
110	254.0	128.0	705	960	1,255	1,588	1,961	2,366	2,820	3,840	5,010	6,340	7,840	9,470	11,300
115	265.5	130.9	720	980	1,282	1,621	2,005	2,420	2,885	3,930	5,120	6,490	8,010	9,680	11,550
120	277.1	133.7	736	1,002	1,310	1,659	2,050	2,470	2,945	4,015	5,225	6,630	8,180	9,900	11,800
130	300.2	139.1	767	1,043	1,365	1,726	2,132	2,575	3,070	4,175	5,450	6,900	8,530	10,300	12,290
140	323.3	144.3	795	1,082	1,415	1,790	2,212	2,650	3,180	4,330	5,650	7,160	8,850	10,690	12,730
150	346.4	149.5	824	1,120	1,466	1,853	2,290	2,760	3,295	4,485	5,850	7,410	9,150	11,070	13,200
175	404.1	161.4	890	1,210	1,582	2,000	2,473	2,985	3,560	4,840	6,310	8,000	9,890	11,940	14,250
200	461.9	172.6	950	1,294	1,691	2,140	2,645	3,190	3,800	5,175	6,760	8,550	10,580	12,770	15,220
250	577.4	193.0	1,063	1,447	1,891	2,392	2,955	3,570	4,250	5,795	7,550	9,570	11,820	14,290	17,020
300	692.8	211.2	1,163	1,582	2,070	2,615	3,235	3,900	4,650	6,330	8,260	10,480	12,940	15,620	18,610

NOTE: The actual quantity discharged by a nozzle will be less than shown in this table. A well-tapered smooth nozzle may be assumed to give 97% to 99% of the values in the table.

*Where there is both an upstream and downstream pressure, the head is a differential head.

Rates of Flow for Certain Plumbing, Household, and Farm Fixtures

Location	Flow Pressure*		Flow Rate	
	psi	*(kPa)*	*gpm*	*(L/min)*
Ordinary basin faucet	8	(55)	2.0	(7.5)
Self-closing basin faucet	8	(55)	2.5	(9.5)
Sink faucet, 3/8 in. (10 mm)	8	(55)	4.5	(17.0)
Sink faucet, 1/2 in. (13 mm)	8	(55)	4.5	(17.0)
Bathtub faucet	8	(55)	6.0	(23.0)
Laundry tub faucet, 1/2 in. (13 mm)	8	(55)	5.0	(19.0)
Faucets per Energy Policy Act of 1992	—	—	≤2.5	—
Shower	8	(55)	5.0	(19.0)
Showers per Energy Policy Act of 1992	—	—	≤2.5	—
Ball-cock for toilet	8	(55)	3.0	(11.0)
Flush valve for toilet	15	(103)	15.0–40.0	(57.0–151.0)†
Toilets gal or L per flush per Energy Policy Act of 1992	—	—	≤1.6	(≤6.1)
Flushometer valve for urinal	15	(103)	15.0	(57.0)
Garden hose, 50 ft (15 m) (3/4-in. [13-mm] sill cock)	30	(207)	5.0	(19.0)
Garden hose, 50 ft (15 m) (5/8-in. [16-mm] outlet)	15	(103)	3.33	(13.0)
Drinking fountain	15	(103)	0.75	(3.0)
Fire hose, 1½ in. (6 mm) (1/2-in. [13-mm] nozzle)	30	(207)	40.0	(151.0)

* Flow pressure is the pressure in the supply near the faucet or water outlet while the faucet or water outlet is wide open and flowing. Flow pressure is measured in pounds per square inch (kilopascals).

† Wide range because designs and types of toilet flush valves vary.

Flow

Wastewater Treatment

Wastewater treatment is a biological system that must be kept in balance. It is a scientific art requiring knowledge of multiple disciplines. New technologies are making treatment more complex as greater regulatory demands are required for the industry.

Unit Processes for Wastewater Reclamation

Constituent	Primary Treatment	Activated Sludge	Nitrification/ Denitrification	Trickling Filter	Rotation Biological Contactor	Coagulation/ Flocculation/ Sedimentation	Filtration After Activated Sludge	Carbon Adsorption	Reverse Osmosis	Ozonation	Chlorination	Ultraviolet
Total suspended solids	++	+++	+++	+++	+++	+++	+++	++	+++			
Total dissolved solids									+++			
Turbidity	++	++	++	++		+++	++	+++	+++	+++		
Color	+	++	++	+	+++	+++	++	+++	+++	+		
Biological oxygen demand	++	+++	+++	+++	+++	+++	++	+++	+++			
Chemical oxygen demand	++	+++	+++	+++		+++	++	++	+++	+++		
Total organic carbon	++	+++	+++	++		+++	++	+++	+++	+++		
Phosphorus	+	++	+++			+++	+++	+++	+++			
NH$_3$–N	+	+++	+++		+++	+	++	++	+++			
NO$_3$–N			++				++	+				

Table continued on next page

Unit Processes for Wastewater Reclamation (continued)

Constituent	Unit Process											
	Primary Treatment	Activated Sludge	Nitrification/ Denitrification	Trickling Filter	Rotation Biological Contactor	Coagulation/ Flocculation/ Sedimentation	Filtration After Activated Sludge	Carbon Adsorption	Reverse Osmosis	Ozonation	Chlorination	Ultraviolet
Cadmium	++	+++	+++	+	++	+++	++	+				
Copper	++	+++	+++	+++	+++	+++	+	++				
Iron	++	+++	+++	++	+++	+++	+++	+++				
Lead	+++	+++	+++	++	+++	+++	+	++				
Zinc	++	++	+++	+++	+++	+++		++				
Foaming agents		+++	+++	+++		++		+++		+		
Total coliform	++	+++	+++	+		+++		+++		+++	+++	+++

NOTES:

+++ = good removals >50%.

++ = intermediate 25%–50%.

+ = low 25%.

Blank cells denote no or inconclusive data, or an increase.

263

Wastewater Treatment

Weir Overflow for Rectangular Clarifier

$$\text{detention time} = \frac{\text{volume of tank}}{\text{flow rate}}$$

$$\text{surface overflow rate} = \frac{\text{flow, gpd}}{\text{tank surface, ft}^2}$$

$$\text{weir overflow rate} = \frac{\text{flow, gpd}}{\text{weir length, ft}}$$

Calculations for Pounds of Biological Oxygen Demand (BOD) and Suspended Solids Loading in a Primary Clarifier

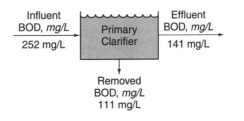

Influent
BOD, *mg/L*
252 mg/L

Primary Clarifier

Effluent
BOD, *mg/L*
141 mg/L

Removed
BOD, *mg/L*
111 mg/L

$$\frac{\text{solids}}{\text{loading}} = \frac{(\text{flow, mgd} \times 8.34 \times \text{MLSS, mg/L})\binom{\text{solids applied,}}{\text{lb/day}}}{\text{surface area, ft}^2\,(0.785 \times D^2)}$$

Where:

MLSS = mixed liquor suspended solids

Filters

$$\text{hydraulic loading rate} = \frac{\text{flow, mgd} \times 8.34 \times \text{BOD, mg/L}}{\text{ft}^2}$$

$$\text{recirculation flow ratio} = \frac{\text{recirculation flow, mgd}}{\text{primary effluent flow, mgd}}$$

Contactors

$$\text{hydraulic loading rate} = \frac{\text{total flow applied, gpd}}{\text{area, ft}^2}$$

$$\text{organic loading rate} = \frac{\text{flow, mgd} \times 8.34 \times \text{soluble BOD, mg/L}}{\text{media area, 1,000 ft}^2}$$

Ponds

$$\text{hydraulic loading rate, gpd/ft}^2 = \frac{\text{flow, gpd}}{\text{area, ft}^2}$$

$$\text{hydraulic loading rate, acre-ft/day/acre} = \frac{\text{flow, acre-ft/day}}{\text{area, acre}}$$

$$\text{BOD, lb} = \text{flow} \times 8.34 \text{ lb/gal} \times \text{mg/L}$$

$$\frac{\text{\% BOD}}{\text{removal}} = \frac{\text{BOD influent, mg/L} - \text{BOD effluent, mg/L}}{\text{BOD influent, mg/L}} \times 100$$

$$\frac{\text{organic loading rate,}}{\text{lb BOD/day/acre}} = \frac{\text{flow, mgd} \times 8.34 \text{ lb/gal} \times \text{BOD, mg/L}}{\text{acre}}$$

$$\text{detention time, days} = \frac{\text{volume of pond, gal}}{\text{flow rate, gpd}}$$

BOD

$$\frac{\text{initial dissolved oxygen (DO), mg/L} - \text{final DO, mg/L}}{\text{sample volume, mL/bottle volume, mL}}$$

Wastewater Treatment

Filter Loading Rate

$$\text{filter loading rate} = \frac{\text{flow, gpm}}{\text{filter area, ft}^2}$$

$$\text{filter loading rate} = \frac{\text{inches of water fall}}{\text{minute}}$$

Filter Backwash Rate

$$\text{filter backwash rate} = \frac{\text{flow, gpm}}{\text{filter area, ft}^2}$$

$$\text{filter backwash rate} = \frac{\text{inches of water rise}}{\text{minute}}$$

Force

$$\text{force} = \text{pressure} \times \text{area}$$

Head

$$\text{head} = \frac{\text{ft-lb}}{\text{lb}}$$

$$\text{velocity head} = \frac{V^2}{64.4 \text{ ft/sec}^2}$$

$$\text{equivalent flow rate} = \frac{\text{actual flow rate} \times 100}{C \text{ value}}$$

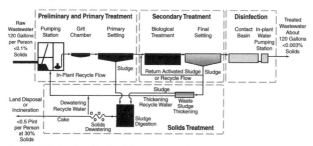

NOTE: Liquid treatment consists of preliminary treatment, primary sedimentation, biological treatment, final sedimentation, and disinfection prior to discharge. Solids treatment consists of digestion of primary and thickened secondary solids, mechanical dewatering, and land disposal.

Courtesy of Pearson Education, Inc.

Conventional Municipal Wastewater Treatment Plant

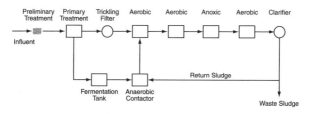

Source: Water and Wastewater Calculations Manual, *copyright 2001, The McGraw-Hill Companies.*

OWASA Nitrification Process

Wastewater Treatment

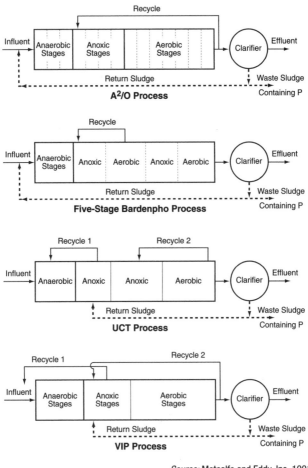

Source: Metcalfe and Eddy, Inc. 1991.

Combined Biological Nitrogen and Phosphorus Removal Processes

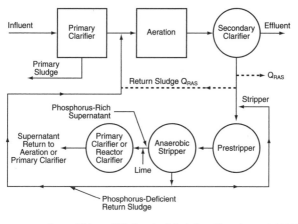

Source: Water and Wastewater Calculations Manual, *copyright 2001,*
The McGraw-Hill Companies.

PhoStrip II Process for Phosphorus and Nitrogen Removal

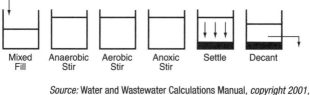

Source: Water and Wastewater Calculations Manual, *copyright 2001,*
The McGraw-Hill Companies.

**Sequencing Batch Reactor for Carbon Oxidation Plus Phosphorus and
Nitrogen Removal**

Wastewater Treatment

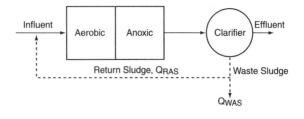

Wuhrmann Process for Nitrogen Removal

Chemicals Used in Wastewater Treatment

Lime—calcium oxide, CaO	Produces calcium carbonate in wastewater which acts as a coagulant for hardness and particulate matter. Often used in conjunction with other coagulants, because by itself, large quantities of lime are required for effectiveness, and lime typically generates more sludge than other coagulants.
Ferrous sulfate— $Fe(SO_4)_3$	Typically used with lime to soften water. The chemical combination forms calcium sulfate and ferric hydroxide. Wastewater must contain dissolved oxygen for reaction to proceed successfully.
Alum or filter alum— $Al_2(SO_4)_3 \bullet 14H_2O$	Used for water softening and phosphate removal. Reacts with available alkalinity (carbonate, bicarbonate, and hydroxide) or phosphate to form insoluble aluminum salts.
Ferric chloride— $FeCl_3$	Reacts with alkalinity or phosphates to form insoluble iron salts.
Polymer	High-molecular-weight compounds (usually synthetic) which can be anionic, cationic, or nonionic. When added to wastewater, can be used for charge neutralization for emulsion-breaking, or as bridge-making coagulants, or both. Can also be used as filter aids and sludge conditioners.

Commercial Forms of Chemical Precipitation Chemicals

Chemical	Commercial Characteristic
Alum	Alum is an off-white crystal that, when dissolved in water, produces acidic conditions. As a solid, alum may be supplied in lumps but is available in ground, rice, or powdered form. Shipments range from 100-lb bags to bulk quantities of 4,000 lb. In liquid form, alum is commonly supplied as a 50% solution delivered in minimum loads of 4,000 gal. The choice between liquid and dry alum depends on the availability of storage space, the method of feeding, and economics.
$FeCl_3$	Ferric chloride, or $FeCl_3$, is available in either dry (hydrate or anhydrous) or liquid form. The liquid form is usually 35%–45% $FeCl_3$. Because higher concentrations of $FeCl_3$ have higher freezing points, lower concentrations are supplied during the winter. It is highly corrosive.
Lime	Lime can be purchased in many forms, with quicklime (CaO) and hydrated lime ($Ca(OH)_2$) being the most prevalent forms. In either case, lime is usually purchased in the dry state, in bags, or in bulk.
Polymer	Polymers may be supplied as a prepared stock solution ready for addition to the treatment process or as a dry powder. Many competing polymer formulations with differing characteristics are available, requiring somewhat different handling procedures. Manufacturers should be consulted for recommended practices and use.

Wastewater Treatment

Approximate Nutrient Composition of Average Sanitary Wastewater Based on 120 gpcd (450 L/person·day)

Parameter	Raw	After Settling	Biologically Treated
Organic content, *mg/L*			
Suspended solids	240	120	30
Biochemical oxygen demand	200	130	30
Nitrogen content, *mg/L as N*			
Inorganic nitrogen	22	22	24
Organic nitrogen	13	8	2
Total nitrogen	35	30	26
Phosphorus content, *mg/L as P*			
Inorganic phosphorus	4	4	3
Organic phosphorus	3	2	2
Total phosphorus	7	6	5

Courtesy of Pearson Education, Inc.

Approximate Composition of Average Sanitary Wastewater (mg/L) Based on 120 gpcd (450 L/person·day)

Parameter	Raw	After Settling	Biological Treated
Total solids	800	680	530
Total volatile solids	440	340	220
Suspended solids	240	120	30
Volatile suspended solids	180	100	20
Biochemical oxygen demand	200	130	30
Inorganic nitrogen as N	22	22	24
Total nitrogen as N	35	30	26
Soluble phosphorus as P	4	4	4
Total phosphorus as P	7	6	5

Courtesy of Pearson Education, Inc.

Grit

The volume of grit removed using a vortex grit unit can be calculated as follows:

$$\text{grit, lb/mgd} = \left(\frac{\text{peak flow, mgd}}{\text{average flow, mgd}} - 1 \right) \times 670$$

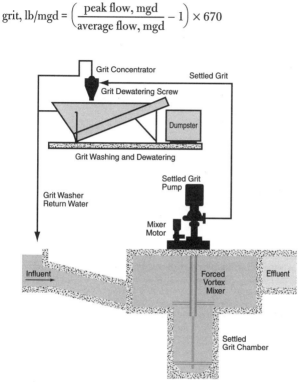

NOTE: The vortex suspends organic solids while grit settles in the lower chamber. The grit pump removes settled grit to be dewatered and held in a dumpster prior to disposal in a landfill.

Courtesy of Pearson Education Inc.

Forced Vortex Unit for Removing Grit

Wastewater Treatment

Typical Design Criteria for Primary Clarifiers

	Average Monthly Flow	Peak Flow
Overflow rates, gpd/ft^2		
USEPA	800–1,200	2,000–3,000
GLUMRB*	1,000	1,500
USEPA with secondary solids	600–800	1,200–1,500
Side water depth, ft		
USEPA	10–13	
GLUMRB	7	
USEPA with secondary solids	13–16	
Weir loading, gpd/ft		
USEPA	10,000–40,000	
GLUMRB	10,000	

Courtesy of Pearson Education, Inc.

* GLUMRB = Great Lakes–Upper Mississippi River Board of State Public Health and Environmental Managers.

Typical Design Parameters for Primary Clarifiers

Type of Treatment	Source	Surface Settling Rate, $m^3/m^2 \cdot day$ (gal/day·ft^2)		Depth, m (ft)
		Average	Peak	
Primary settling followed by secondary treatment	USEPA 1975a	33–49 (800–1,200)	81–122 (2,000–3,000)	3–3.7 (10–12)
	GLUMRB–Ten States Standards and Illinois EPA 1998	600		Minimum 2.1 (7)
Primary settling with waste-activated sludge return	USEPA 1975a	24–33 (600–800)	49–61 (1,200–1,500)	3.7–4.6 (12–15)
	Ten States Standards, GLUMRB 1996	≤41 (≤1,000)	≤61 (≤1,500)	Minimum 3.0 (10)

Source: Water and Wastewater Calculations Manual, *copyright 2001, The McGraw-Hill Companies.*

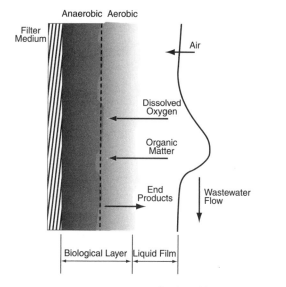

Courtesy of Pearson Education, Inc.

Biological Process in a Filter Bed

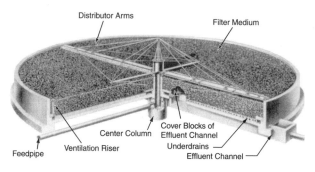

Courtesy of Pearson Education, Inc.

Cut-Away View of Stone-Media Trickling Filter With Concrete Side Walls

Wastewater Treatment

$$\text{biological oxygen demand (BOD) loading} = \frac{\text{settled wastewater BOD}}{\text{volume of filter media}}$$

Where:

BOD loading = pounds of BOD applied per 1,000 ft³/day (g/m³·day)

settled BOD = wastewater BOD remaining after primary sedimentation, in lb/day (g/day)

volume of media = volume of stone in the filters, in thousands of ft³ (m³)

$$\text{hydraulic loading} = \frac{Q + Q_R}{A}$$

Where:

hydraulic loading = mil gal/acre/day (m³/m²·day)

Q = wastewater flow, in mgd (m³/day)

Q_R = recirculation flow, in mgd (m³/day)

A = surface area of filters, in acres (m²)

$$R = \frac{Q_R}{Q}$$

Where:

R = recirculation ratio

Q_R and Q = (same as above)

Typical Loadings for Trickling Filters With a 5-to-7-ft Depth of Stone or Slag Media

	High Rate	Two Stage
Biological oxygen demand loading		
lb/1,000 ft³·day*	30–90	45–70
lb/acre-ft·day	1,300–3,900	2,000–3,000
Hydraulic loading		
mil gal/acre·day†	10–30	10–30
gpm/ft²	0.16–0.48	0.16–0.48
Recirculation ratio	0.5–3.0	0.5–4.0

Courtesy of Pearson Education, Inc.

* 1.0 lb/1,000 ft³·day = 16.0 g/m³·day.
† 1.0 mil gal/acre·day = 0.935 m³/m²·day.

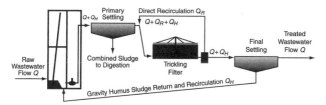

Profile of a single-stage trickling filter showing related wastewater flow diagrams including in-plant recirculation

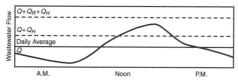

General flow patterns: Q = wastewater influent flow; $Q + Q_H$ = influent plus humus return from the bottom of the clarifier; and $Q + Q_R + Q_H$ = flow to the filter with direction and indirect recirculation

Courtesy of Pearson Education, Inc.

Single-Stage Trickling Filter Plant

Wastewater Treatment

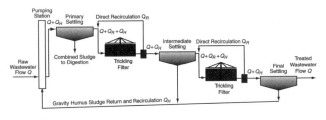

Courtesy of Pearson Education, Inc.

Typical Flow Diagram for a Two-Stage Trickling Filter Plant With Intermediate Clarifier

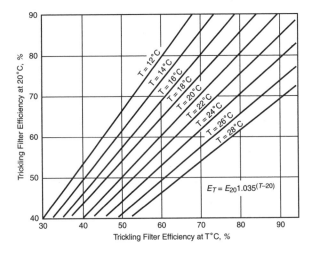

$$E_T = E_{20}1.035^{(T-20)}$$

Courtesy of Pearson Education, Inc.

Diagram for Correcting Biological Oxygen Demand Removal Efficiency From National Research Council Data at 20°C to Efficiency at Other Temperatures Between 12°C and 28°C

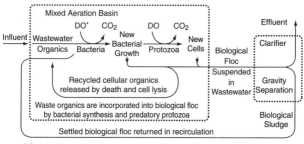

*DO = dissolved oxygen.

Generalized Biological Process in Aeration (Activated-Sludge) Treatment

The following equation calculates the *F/M* value as BOD applied/day/unit mass of MLSS in the aeration tank:

$$\frac{F}{M} = \frac{Q \times \text{BOD}}{V \times \text{MLSS}}$$

Where:

F/M = food-to-microorganism ratio, in lb BOD/day per lb MLSS (g BOD/day per g MLSS)

Q = wastewater flow, in mgd (m³/d)

BOD = wastewater BOD, in mg/L (g/m³)

V = liquid volume of aeration tank, in mil gal (m³)

MLSS = mixed liquor suspended solids in the aeration basin, in mg/L (g/m³)

Wastewater Treatment

The following equation calculates sludge age on the basis of the mass of MLSS in the aeration tank relative to the mass of suspended solids in the wastewater effluent and waste sludge:

$$\text{sludge age} = \frac{\text{MLSS} \times V}{\text{SS}_e \times Q_e + \text{SS}_w \times W_w}$$

Where:

sludge age = mean cell residence time, in days

MLSS = mixed liquor suspended solids, in mg/L (g/m^3)

V = volume of aeration tank, in mil gal (m^3)

SS$_e$ = suspended solids in wastewater effluent, in mg/L (g/m^3)

Q_e = quantity of wastewater effluent, in mgd (m^3/day)

SS$_w$ = suspended solids in waste sludge, in mg/L (g/m^3)

Q_w = quantity of waste sludge, in mgd (m^3/day)

Summary of Loadings and Operational Parameters for Aeration Processes

Process	Biological Oxygen Demand (BOD) Loading, lb BOD/day per 1,000 ft³	Mixed Liquor Suspended Solids (MLSS), mg/L	Food-to-Microorganism (F/M) Ratio, lb BOD/day per lb MLSS	Sludge Age, days	Aeration Period, hr	Return Sludge Rates, %	BOD Removal Efficiency, %
Conventional	20–40	1,000–3,000	0.2–0.5	5–15	4.0–7.5	20–40	80–90
Step aeration	40–60	1,500–3,500	0.2–0.5	5–15	4.0–7.0	30–50	80–90
Extended aeration	10–20	2,000–8,000	0.05–0.2	≥20	20–30	50–100	85–95
High-purity oxygen	≥120	4,000–8,000	0.6–1.5	3–10	1.0–3.0	30–50	80–90

Courtesy of Pearson Education, Inc.

1.0 lb/1,000 ft³·day = 16.0 g/m³·day.

1.0 lb/day/lb MLSS = 1.0 g/day·g/MLSS.

Wastewater Treatment

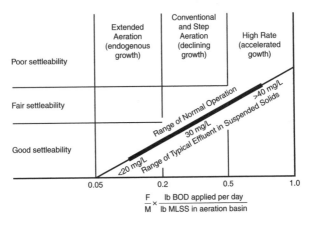

Courtesy of Pearson Education, Inc.

Approximate Relationship Between Activated-Sludge Settleability and Operating Food-to-Microorganism Ratio

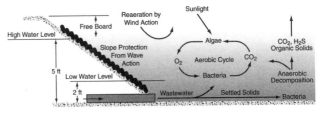

Courtesy of Pearson Education, Inc.

Facultative Stabilization Pond Showing the Basic Biological Reactions of Bacteria and Algae

Minimum National Performance Standards for Publicly Owned Treatment Works (Secondary Treatment and Its Equivalency)

Parameter	30-Day Average Shall Not Exceed	7-Day Average Shall Not Exceed
Conventional Secondary Treatment Processes		
5-day biochemical oxygen demand,* BOD_5		
Effluent, *mg/L*	30	45
Percent removal†	85	—
5-day carbonaceous BOD,* $CBOD_5$		
Effluent, *mg/L*	25	40
Percent removal†	85	—
Suspended solids		
Effluent, *mg/L*	30	45
Percent removal†	85	—
pH	6.0–9.0 at all times	—
Whole effluent toxicity	Site specific	—
Fecal coliform, *MPN*‡*/100 mL*	200	400
Stabilization Ponds and Other Equivalent of Secondary Treatment		
5-day biochemical oxygen demand,* BOD_5		
Effluent, *mg/L*	45	65
Percent removal†	65	—
5-day carbonaceous BOD,* $CBOD_5$		
Effluent, *mg/L*	40	60
Percent removal†	65	—
Suspended solids		
Effluent, *mg/L*	45	65
Percent removal†	65	—
pH	6.0–9.0 at all times	—
Whole effluent toxicity	Site specific	—
Fecal coliform, *MPN/100 mL*	200	400

* Chemical oxygen demand (COD) or total organic carbon (TOC) may be substituted for BOD_5 when a long-term BOD_5:COD or BOD_5:TOC correlation has been demonstrated.

† Percent removal may be waived on a case-by-case basis for combined sewer service areas and for separated sewer areas not subject to excessive inflow and infiltration (I/I) where the base flow plus infiltration is ≤120 gpcd and the base flow plus I/I is ≤275 gpcd.

‡ MPN = most probable number.

Wastewater Treatment

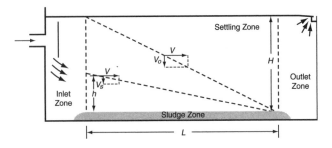

Discrete Particle Settling in an Ideal Settling Tank

The flow rate of wastewater is

$$V_O = Q/A = Q/WL = \frac{g(\rho_s - \rho)d^2}{18\mu}$$

Where:

Q = flow, in gpd (m^3/day)

A = surface area of the settling zone, in ft^2 (m^2)

V_O = overflow rate or surface loading rate, in gal/(ft^2·day) (m^3/[m^2·day])

W, L = width and length of the tank, in ft (m)

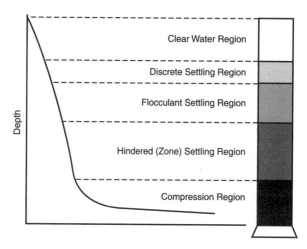

Source: Water and Wastewater Calculations Manual, *copyright 2001,*
The McGraw-Hill Companies.

Settling Regions for Concentrated Suspensions

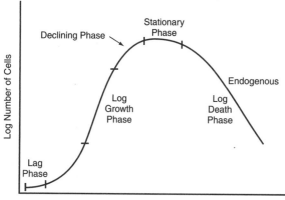

Source: Water and Wastewater Calculations Manual, *copyright 2001,*
The McGraw-Hill Companies.

Bacterial Density With Growth Time

Wastewater Treatment

Guidelines for Return-Activated Sludge Flow Rate

Type of Process	Percent of Design Average Flow	
	Minimum	Maximum
Conventional	15	100
Carbonaceous stage of separate-stage nitrification	15	100
Step-feed aeration	15	100
Complete-mix	15	100
Contact stabilization	50	150
Extended aeration	50	150
Nitrification stage of separate-stage nitrification	50	200

Typical Design Parameters for Secondary Sedimentation Tanks

Type of Treatment	Hydraulic Loading, $gal/(day \cdot ft^2)^*$		Solids Loading,[†] $lb\ solids/(day \cdot ft^2)^*$		Depth, ft
	Average	Peak	Average	Peak	
Settling following tracking filtration	400–600	1,000–2,000	0	0	10–12
Settling following air-activated sludge (excluding extended aeration)	400–800	1,000–1,200	20–30	50	12–15
Settling following extended aeration	200–400	800	20–30	50	12–15
Settling following oxygen-activated sludge with primary settling	400–800	1,000–1,200	25–35	50	12–15

* $gal/(day \cdot ft^2) \times 0.0407 = m^3/(m^2 \cdot day)$; $lb/(day \cdot ft^2) \times 4.883 = kg/(day \cdot m^2)$
† Allowable solids loading area generally governed by sludge thickening characteristics associated with cold weather operations.

Recommended Design Overflow Rate and Peak Solids Loading Rate for Secondary Settling Tanks Following Activated-Sludge Processes

Treatment Process	Surface Loading at Design Peak Hourly Flow,[*] gal/d·ft^2 (m^3/m^2·day)	Peak Solids Loading Rate,[†] lb/d·ft^2 (kg/(d·m^2))
Conventional	1,200 (49)	50 (244)
Step aeration	or	
Complete mix	1,000 (41)	
Contact stabilization		
Carbonaceous stage of separate-stage nitrification		
Extended aeration	1,000 (41)[‡]	35 (171)
Single-stage nitrification		
Two-stage nitrification	800 (33)	35 (171)
Activated sludge with chemical addition to mixed liquor for phosphorus removal	900 (37)[§]	As above

[*] Based on influent flow only.

[†] For plant effluent TSS ≤20 mg/L.

[‡] Computed on the basis of design maximum daily flow rate plus design maximum return sludge rate requirements, and the design MLSS under aeration.

[§] When effluent P concentration of 1.0 mg/L or less is required.

Wastewater Treatment

Recommended Chlorine Dosing Capacity for Various Types of Treatment Based on Design Average Flow

Type of Treatment	Illinois EPA Dosage, mg/L	GLUMRB* Dosage, mg/L
Primary settled effluent	20	
Lagoon effluent (unfiltered)	20	
Lagoon effluent (filtered)	10	
Trickling filter plant effluent	10	10
Activated sludge plant effluent	6	8
Activated sludge plant with chemical addition	4	
Nitrified effluent		6
Filtered effluent following mechanical biological treatment	4	6

* GLUMRB = Great Lakes–Upper Mississippi River Board of State Public Health and Environmental Managers.

DIFFUSERS

Some advantages and disadvantages of various fine pore diffusers are listed in the following sections.

Advantages

- Exhibit high oxygen-transfer efficiencies
- Exhibit high aeration efficiencies (mass oxygen transferred per unit power per unit time)
- Can satisfy high oxygen demands
- Are easily adaptable to existing basins for plant upgrades
- Result in lower volatile organic compound emissions than nonporous diffusers or mechanical aeration devices

Disadvantages

- Fine pore diffusers are susceptible to chemical or biological fouling, which may impair transfer efficiency and generate high head loss. As a result, they require routine cleaning. (Although not totally without cost, cleaning does not need to be expensive or troublesome.)

- Fine pore diffusers may be susceptible to chemical attack (especially perforated membranes). Therefore, care must be exercised in the proper selection of materials for a given wastewater.

- Because of the high efficiencies of fine pore diffusers at low airflow rates, airflow distribution is critical to their performance, and selection of proper airflow control orifices is important.

- Because of the high efficiencies of fine pore diffusers, required airflow in an aeration basin (normally at the effluent end) may be dictated by mixing, not oxygen transfer.

- Aeration basin design must incorporate a means to easily dewater the tank for cleaning. In small systems where no redundancy of aeration tanks exists, an in situ, non-process–interruptive method of cleaning must be considered.

SEQUENCING BATCH REACTORS

Some advantages and disadvantages of sequencing batch reactors (SBRs) are listed in the following sections.

Advantages

- Equalization, primary clarification (in most cases), biological treatment, and secondary clarification can be achieved in a single reactor vessel.
- Operating flexibility and control
- Minimal footprint
- Potential capital cost savings by eliminating clarifiers and other equipment

Disadvantages

- A higher level of sophistication of timing units and controls is required (compared to conventional systems), especially for larger systems
- Higher level of maintenance (compared to conventional systems) associated with more sophisticated controls, automated switches, and automated valves

Wastewater Treatment

- Potential of discharging floating or settled sludge during the draw or decant phase with some SBR configurations
- Potential plugging of aeration devices during selected operating cycles, depending on the aeration system used by the manufacturer
- Potential requirement for equalization after the SBR, depending on the downstream processes

SBR manufacturers will typically provide a process guarantee to produce an effluent of less than

- 10 mg/L biological oxygen demand
- 10 mg/L total suspended solids
- 5–8 mg/L total nitrogen
- 1–2 mg/L total phosphorus

Key Design Parameters for a Conventional Load

Parameter	Municipal	Industrial
Food to mass (F:M)	0.15–0.4/day	0.15–0.6/day
Treatment cycle duration	4 hours	4–24 hours
Typically low water level mixed liquor suspended solids	2,000–2,500 mg/L	2,000–4,000 mg/L
Hydraulic retention time	6–14 hours	Varies

Case Studies for Several SBR Installations

Flow, mgd	Reactors			Blowers	
	No.	Size, *ft*	Volume, *mil gal*	No.	Size, *hp*
0.012	1	18 × 12	0.021	1	15
0.10	2	24 × 24	0.069	3	7.5
1.2	2	80 × 80	0.908	3	125
1.0	2	58 × 58	0.479	3	40
1.4	2	69 × 69	0.678	3	60
1.46	2	78 × 78	0.910	4	40
2.0	2	82 × 82	0.958	3	75
4.25	4	104 × 80	1.556	5	200
5.2	4	87 × 87	1.359	5	125

Source: Courtesy of Aqua-Aerobic Systems, Inc.
NOTE: These case studies and sizing estimates are site specific to individual treatment systems.

Installed Cost per Gallon of Wastewater Treated

Design Flow Rate, *mgd*	Budget Level Equipment Cost, *$/gal*
0.5–1.0	1.96–5.00
1.1–1.5	1.83–2.69
1.5–2.0	1.65–3.29

Source: Courtesy of Aqua-Aerobic Systems, Inc.

Wastewater Treatment

INTERMITTENT SAND FILTERS _____

Typical Design Criteria for Intermittent Sand Filters

Item	Design Criteria
Pretreatment	Minimum level: septic tank or equivalent
Filter medium	
Material	Washed durable granular material
Effective size	0.25–0.75 mm
Uniformity coefficient	<4.0
Depth	18–36 in.
Underdrains	
Type	Slotted or perforated pipe
Slope	0%–0.1%
Size	3–4 in.
Hydraulic loading	2–5 gal/ft^2·day
Organic loading	0.0005–0.002 lb/ft^2·day
Pressure distribution	
Pipe size	1–2 in.
Orifice size	1/8–1/4 in.
Head on orifice	3–6 ft
Lateral spacing	1–4 ft
Orifice spacing	1–4 ft
Dosing	
Frequency	12–48 times/day
Volume/orifice	0.15–0.30 gal/orifice/dose
Dosing tank volume	0.5–1.5 flow/day

Some advantages and disadvantages of intermittent sand filters (ISFs) are listed in the following sections.

Advantages

- ISFs produce a high-quality effluent that can be used for drip irrigation or can be surface-discharged after disinfection.
- Drainfields can be small and shallow.
- ISFs have low-energy requirements.
- ISFs are easily accessible for monitoring and do not require skilled personnel to operate.
- No chemicals are required.
- If sand is not feasible, other suitable media can be substituted and may be found locally.
- Construction costs for ISFs are moderately low, and the labor is mostly manual.
- The treatment capacity can be expanded through modular design.
- ISFs can be installed to blend into the surrounding landscape.

Disadvantages

- The land area required may be a limiting factor.
- Regular (but minimal) maintenance is required.
- Odor problems could result from open-filter configurations and may require buffer zones from inhabited areas.
- If appropriate filter media are not available locally, costs could be higher.
- Clogging of the filter media is possible.

SEPTAGE

Some advantages and disadvantages of septage are listed in the following sections.

Advantages

Use of treatment plants provides regional solutions to septage management.

Disadvantages

- May need a holding facility during periods of frozen or saturated soil.
- Need a relatively large, remote land area for the setup of the septic system.
- Capital and operation and maintenance costs tend to be high.
- Some limitations to certain management options of untreated septage include lack of available sites and potential odor and pathogen problems. These problems can be reduced by pretreating and stabilizing the septage before it is applied to the land.
- Septage treated at a wastewater treatment facility has the potential to upset processes if the septage addition is not properly regulated.

Characteristics of Septage Conventional Parameters[*]

Parameter	Concentration	
	Minimum	**Maximum**
Total solids	1,132	130,475
Total volatile solids	353	71,402
Total suspended solids	310	93,378
Volatile suspended solids	95	51,500
Biochemical oxygen demand	440	78,600
Chemical oxygen demand	1,500	703,000
Total Kjeldahl nitrogen	66	1,060
Ammonia nitrogen	3	116
Total phosphorus	20	760
Alkalinity	522	4,190
Grease	208	23,368
pH	1.5	12.6
Total coliform	10^7/100 mL	10^9/100 mL
Fecal coliform	10^6/100 mL	10^8/100 mL

[*] Measurements are in milligrams per liter unless otherwise indicated.

Wastewater Treatment

Sources of Septage

Description Rate	Removal Pump-out	Characteristics
Septic tank	2–6 years, but can vary with location and local ordinances	Concentrated BOD, solids, nutrients, variable toxins (such as metals), inorganics (sand), odor, pathogens, oil, and grease
Cesspool	2–10 years	Concentrated BOD, solids, nutrients, variable toxins, inorganics, sometimes high grit, odor, pathogens, oil, and grease
Privies/portable toilets	1 week to months	Variable BOD, solids, inorganics, odor, pathogens, and some chemicals
Aerobic tanks	Months to 1 year	Variable BOD, inorganics, odor, pathogens, and concentrated solids
Holding tanks (septic tank with no drainfield, typically a local requirement)	Days to weeks	Variable BOD, solids, inorganics, odor, and pathogens; similar to raw wastewater solids
Dry pits (associated with septic fields)	2–6 years	Variable BOD, solids, inorganics, and odor
Miscellaneous—may exhibit characteristics of septage		
Private wastewater treatment plants	Variable	Septic tank
Boat pump-out station	Variable	Portable toilets
Grit traps	Variable	Oil, grease, solids, inorganics, odor, and variable BOD
Grease traps	Weeks to months	Oil, grease, BOD, viscous solids, and odor

Courtesy of Water Environment Federation.

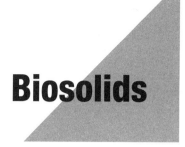

Biosolids

At the end of every wastewater system is the residue of the process, the biosolids. Disposal of biosolids is becoming an environmental concern. New treatments, disinfection processes, and disposal methods are available to help systems comply with increased regulations.

SLUDGE PROCESSING CALCULATIONS_____

Percent Solids and Sludge Pumping

The two basic equations used to calculate percent solids are

$$\% \text{ solids} = \frac{\text{total solids, g}}{\text{sludge sample, g}} \times 100$$

$$\% \text{ solids} = \frac{\text{solids, lb/day}}{\text{sludge, lb/day}} \times 100$$

The basic equation for sludge thickening and sludge volume changes is

lb solids in unthickened sludge = lb solids in thickened sludge

or

$$\left(\begin{array}{c} \text{unthickened} \\ \text{sludge, lb/day} \end{array} \right) (\% \text{ solids}) = \left(\begin{array}{c} \text{thickened} \\ \text{sludge, lb/day} \end{array} \right) (\% \text{ solids})$$

Gravity Thickening

The two basic equations for determining gravity thickening are

$$\text{hydraulic loading rate, gpd/ft}^2 = \frac{\text{flow, gpd}}{\text{area, ft}^2}$$

$$\text{solids loading rate, lb/day/ft}^2 = \frac{\text{solids, lb/day}}{\text{area, ft}^2}$$

If the pounds-per-day solids is not given directly, it can be calculated using pounds-per-day sludge and percent solids. The formula follows.

$$\text{solids loading rate, lb/day/ft}^2 = \frac{\text{solids, lb/day} \times \% \text{ solids}}{\text{area, ft}^2}$$

The basic equation to determine the proper wasting rates for activated sludge processes to maintain a desired food-to-mircoorganism (F/M) ratio is

$$\text{F/M} = \frac{\begin{array}{c} \text{biological oxygen demand} \\ \text{entering the aeration tank, lb} \end{array}}{\begin{array}{c} \text{mixed liquor volatile suspended solids} \\ \text{under aeration, lb} \end{array}}$$

Mean Cell Residence Time

The two basic equations for determining mean cell residence time (MCRT) are

$$\text{MCRT} = \frac{\left(\begin{array}{c}\text{aeration tank}\\\text{suspended solids, lb}\end{array}\right) + \left(\begin{array}{c}\text{clarifier}\\\text{suspended solids, lb}\end{array}\right)}{\left(\begin{array}{c}\text{total suspended solids}\\\text{wastes, lb/day}\end{array}\right) + \left(\begin{array}{c}\text{effluent suspended}\\\text{solids, lb/day}\end{array}\right)}$$

$$\text{MCRT} = \frac{\left(\dfrac{\text{mil gal} \times 8.34 \times \text{mg/L}}{\text{MLSS}}\right) + \left(\begin{array}{c}\text{mil gal} \times 8.34 \times \text{RAS}\\\text{suspended solids}\end{array}\right)}{\left(\begin{array}{c}\text{mgd} \times 8.34 \times \text{mg/L RAS}\\\text{suspended solids}\end{array}\right) + \left(\begin{array}{c}\text{mgd} \times 8.34 \times \text{mg/L}\\\text{suspended solids}\end{array}\right)}$$

Sludge Age

The basic equation for determining sludge age is

$$\text{sludge age, days} = \frac{\text{MLSS, lb}}{\text{suspended solids added, lb/day}}$$

or

$$\begin{array}{c}\text{sludge}\\\text{age,}\\\text{days}\end{array} = \frac{\text{aeration volume, mil gal} \times 8.34 \times \text{MLSS, mg/L}}{\text{mgd} \times 8.34 \times \text{mg/L primary effluent suspended solids}}$$

Vacuum Filter Dewatering

Equations for determining filter loading rates, filter yield, and percent solids recovery are

$$\text{filter loading rate, lb/hr/ft}^2 = \frac{\text{solids to filter, lb/hr}}{\text{surface area, ft}^2}$$

$$\text{filter yield, lb/hr/ft}^2 = \frac{\left(\begin{array}{c}\text{wet cake flow,}\\\text{lb/hr}\end{array}\right)\left(\dfrac{\text{cake, \% solids}}{100}\right)}{\text{filter area, ft}^2}$$

$$\text{\% solids recovery} = \frac{\left(\begin{array}{c}\text{wet cake flow,}\\\text{lb/hr}\end{array}\right)\left(\dfrac{\text{cake, \% solids}}{100}\right)}{\left(\begin{array}{c}\text{sludge feed,}\\\text{lb/day}\end{array}\right)\left(\dfrac{\text{feed, \% solids}}{100}\right)}$$

Biosolids

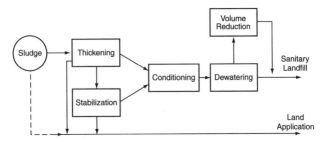

Source: Water and Wastewater Calculations Manual, *copyright 2001,*
The McGraw-Hill Companies.

Sludge Processing Alternatives

Plate and Frame Filter Press Dewatering

Sludge can be dewatered using a plate and frame filter press. It
works by pressing water out of sludge through the use of plates.
Sludge flows in the spaces between the plates and water is pressed
out. The plates are then separated and the cake falls out into a
hopper or onto a conveyor belt.

The equations for determining the solids loading rate and the
net filter yield of a plate and frame filter press are

$$\text{solids loading rate, lb/hr/ft}^2 = \frac{\text{sludge, gph} \times 8.34 \text{ lb.gal} \times \left(\frac{\% \text{ solids}}{100}\right)}{\text{plate area, ft}^2}$$

$$\text{net filter yield, lb/hr/ft}^2 = \frac{\text{lb/hr}}{\text{ft}^2} \times \frac{\text{filtration run time}}{\text{total cycle time}}$$

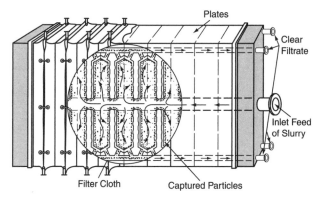

Schematic Cross-Section of a Plate and Frame Filter Press Chamber
Area During Fill Cycle

Belt Filter Press Dewatering

Sludge can be dewatered using a belt filter press. The sludge is
pressed between belts into a cake. The cake is fed into a hopper or
onto a conveyor belt.

The equations for determining the hydraulic loading rate and
the sludge feed rate of a belt filter press are

$$\text{hydraulic loading rate} = \frac{\text{flow, gpm}}{\text{belt width, ft}}$$

$$\text{sludge feed rate} = \frac{\text{sludge fed into press, lb/day}}{\text{operating time, hr/day}}$$

$$\begin{array}{l}\text{volatile}\\ \text{solids, } = \text{ sludge, gpd} \times 8.34 \times \left(\frac{\text{\% solids}}{100}\right) \times \left(\frac{\text{\% volatile solids}}{100}\right)\\ \text{lb/day}\end{array}$$

Digester Loading Rate

Sludge is sent to the digester to stabilize the organic (volatile) portion of the sludge.

$$\begin{array}{l} \text{digester} \\ \text{loading} = \\ \text{rate} \end{array} \dfrac{\text{sludge, gpd} \times 8.34 \times \left(\dfrac{\% \text{ solids}}{100}\right) \times \left(\dfrac{\% \text{ volatile solids}}{100}\right)}{3.14 \times r \times r \times \text{sludge depth, ft}}$$

Volatile Acids/Alkalinity Ratio

The anaerobic digestion process requires an intricate balance between the acid and alkalinity stages. Therefore, by determining the volatile acids/alkalinity ratio, the digestion process can be tracked.

$$\text{volatile acids/alkalinity ratio} = \dfrac{\text{volatile acids, mg/L}}{\text{alkalinity, mg/L}}$$

Digester Gas Production

Gas produced during anaerobic digestion can be used as fuel for heating the digesters and buildings, for driving gas engines, and so forth. The volume of gas produced is an important indicator of the progress of the sludge digestion process.

$$\text{digester gas production} = \dfrac{\text{gas produced, ft}^3/\text{day}}{\text{gpd} \times 8.34 \times \left(\dfrac{\% \text{ solids}}{\text{day}}\right)}$$

$$\times \left(\dfrac{\% \text{ volatile solids}}{100}\right)$$

$$\times \left(\dfrac{\% \text{ volatile solids reduced}}{100}\right)$$

Percent Volatile Solids Reduction

The percent volatile solids reduction is one of the best indicators of the effectiveness of the anaerobic digester process. This reduction can be as high as 70 percent.

$$\% \text{ volatile solids reduction} = \frac{\text{in} - \text{out}}{\text{in} - (\text{in} \times \text{out})} \times 100$$

Settleable Solids

The basic equation for determining settleable solids in milligrams per liter is

$$\frac{(\text{final weight, mg} - \text{initial weight, mg}) \times 1{,}000 \text{ mL/L} \times 1{,}000 \text{ mg/g}}{\text{mL/sec filtered}}$$

Total Solids and Volatile Solids

The basic equations for determining percent total solids, percent volatile solids, and percent fixed matter are

$$\% \text{ total solids} = \frac{\text{mass of dry solids} \ (M3 - M1) \times 100}{\text{mass of wet sludge} \ (M2 - M1)}$$

$$\% \text{ volatile matter} = \frac{\text{mass of volatile solids} \ (M3 - M4) \times 100}{\text{mass of dry solids} \ (M3 - M1)}$$

$$\% \text{ fixed matter} = \frac{\text{mass of fixed matter} \ (M3 - M1) \times 100}{\text{mass of dry solids} \ (M3 - M1)}$$

Where:

All weights are in grams.

  M1 = mass of the dish

  M2 = mass of the dish and wet sample

  M3 = mass of the dish and dry sample

  M4 = mass of the dish and fixed matter

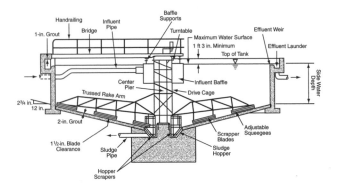

Gravity Thickener

GRAVITY THICKENING

Some advantages and disadvantages of gravity thickening are listed in the following sections.

Advantages

- Gravity thickening equipment is simple to operate and maintain.
- Gravity thickening has lower operating costs than other thickening methods such as dissolved air flotation (DAF), gravity belt, or centrifuge thickening. For example, an efficient gravity thickening operation will save costs incurred in downstream solids handling steps.

In addition, facilities that land-apply liquid biosolids can benefit from thickening in several ways, as follows:

- Truck traffic at the plant and the farm site can be reduced.
- Trucking costs can be reduced.
- Existing storage facilities can hold more days of biosolids production.
- Smaller storage facilities can be used.

- Less time will be required to transfer solids to the applicator vehicle and to incorporate or surface-apply the thickened solids.
- Crop nitrogen demand can be met with fewer passes of the applicator vehicle, reducing soil compaction.

Disadvantages

- Scum buildup can cause odors. This buildup, which can occur because of long retention times, can also increase the torque required in the thickener. Finally, scum buildup is unsightly.
- Grease may build up in the lines and cause a blockage. This can be prevented by quick disposal or a backflush.
- Septic conditions will generate sulfur-based odors. This can be mitigated by minimizing detention times in the collection system and at the plant, or by using oxidizing agents.
- Supernatant does not have solids concentrations as low as those produced by a DAF or centrifuge thickener. Belt thickeners may produce supernatant with lower solids concentrations depending on the equipment and solids characteristics.
- More land area is needed for gravity thickener equipment than for a DAF gravity belt or centrifuge thickener.
- Solids concentrations in the thickened solids are usually lower than for a DAF gravity belt or centrifuge thickener.

Maintenance Checklist

Weekly

- Check all oil levels and ensure the oil fill cap vent is open.
- Check condensation drains and remove any accumulated moisture.
- Examine drive control limit switches.
- Visually examine the skimmer to ensure that it is in proper contact with the scum baffle and the scum box.
- Visually examine instrumentation and clean probes.

Performance of Conventional Gravity Thickening

Type of Solids	Feed, % total solids	Thickened Solids, % total solids
Primary (PRI)	0.6–6	5–10
Trickling filter (TF)	1–4	3–6
Rotating biological contactor (RBC)	1–3.5	2–5
Waste-activated solids (WAS)	0.2–1	2–3
PRI + WAS	3–6	8–15
RPI + TF	2–6	5–9
PRI + RBC	2–6	5–8
PRI + lime	3–4.5	10–15
PRI + (WAS + iron)	1.5	3
PRI + (WAS + aluminum salts)	0.2–0.4	4.5–6.5
Anaerobically digested PRI + WAS	4	8

Adapted with permission from Water Environment Federation (1996) *Operation of Municipal Wastewater Treatment Plants*, 5th ed.; Manual of Practice No. 11; Alexandria, Virginia.

Monthly

- Inspect skimmer wipers for wear.
- Adjust drive chains or belts.

Annually

- Disassemble the drive and examine all gears, oil seals, and bearings.
- Check oil for the presence of metals, which may be a warning sign of future problems.
- Replace any part with an expected life of less than 1 year.

Factors Affecting Gravity Thickening Performance

Factor	Effect
Nature of the solids feed	Affects the thickening process because some solids thicken more easily than others.
Freshness of feed solids	High solids age can result in septic conditions.
High volatile solids concentrations	Hamper gravity settling due to reduced particle specific gravity.
High hydraulic loading rates	Increase velocity and cause turbulence that will disrupt settling and carry the lighter solids past the weirs.
Solids loading rate	If rates are high, there will be insufficient detention time for settling. If rates are too low, septic conditions may arise.
Temperature and variation in temperature of thickener contents	High temperatures will result in septic conditions. Extremely low temperatures will result in lower settling velocities. If temperature varies, settling decreases due to stratification.
High solids blanket depth	Increases the performance of the settling by causing compaction of the lower layers, but it may result in solids being carried over the weir.
Solids residence time	An increase may result in septic conditions. A decrease may result in only partial settling.
Mechanism and rate of solids withdrawal	Must be maintained to produce a smooth and continuous flow. Otherwise, turbulence, septic conditions, altered settling, and other anomalies may occur.
Chemical treatment	Chemicals—such as potassium permanganate, polymers, or ferric chloride—may improve settling and/or supernatant quality.
Presence of bacteriostatic agents or oxidizing agents	Allows for longer detention times before anaerobic conditions create gas bubbles and floating solids.
Cationic polymer addition	Helps thicken waste-activated solids and clarify the supernatant.
Use of metal salt coagulants	Improves overflow clarity but may have little impact on underflow concentration.

Gravity Thickening Troubleshooting Guide

Indicators	Probable Cause	Check or Monitor	Solution
Septic odor, rising solids	Thickened solids pumping rate is too slow; thickener overflow rate is too low.	Check thickened solids pumping system for proper operation; check thickener collection mechanism for proper operation.	Increase pumping rate of thickened solids; increase influent flow to thickener—a portion of the secondary effluent may be pumped to thickener to bring overflow rate to 400–600 gpd/ft^2; chlorinate influent solids.
Thickened solids not thick enough	Overflow rate is too high; thickened solids pumping rate is too high; short-circuiting of flow through tank.	Check overflow rate; use dye or other tracer to check for circulation.	Decrease influent solids flow rate; decrease pumping rate of thickened solids; check effluent weirs and repair or re-level; check influent baffles and repair or relocate.
Torque overload of solids collecting mechanism	Heavy accumulation of solids; foreign object jammed in mechanism; improper alignment of mechanism.	Probe along front of collector arms.	Agitate solids blanket in front of collector arms with water jets; increase solids removal rate; attempt to remove foreign object with grappling device; if problem persists, drain thickener and check mechanism for free operation.

Table continued on next page

308

Gravity Thickening Troubleshooting Guide (continued)

Indicators	Probable Cause	Check or Monitor	Solution
Surging flow	Poor influent pump programming	Pump cycling	Modify pump cycling; reduce flow and increase time.
Excessive biological growths on surfaces and weirs (slimes, etc.)	Inadequate cleaning program		Frequent and thorough cleaning of surfaces; apply chlorination.
Oil leak	Oil seal failure	Oil seal	Replace seal.
Noisy or hot bearing or universal joint	Excessive wear; improper alignment; lack of lubrication	Alignment; lubrication	Replace, lubricate, or align joint or bearing as required.
Pump overload	Improper adjustment of packing; clogged pump	Check packing; check for trash in pump.	Adjust packing; clean pump.
Fine solids particles in effluent	Waste-activated solids	Portion of waste-activated solids (WAS) in thickener effluent	Better conditioning of the WAS portion of the solids; thicken WAS in a flotation thickener.

Adapted with permission from Water Environment Federation (1996) *Operation of Municipal Wastewater Treatment Plants*, 5th ed.; Manual of Practice No. 11; Alexandria, Virginia.

Biosolids

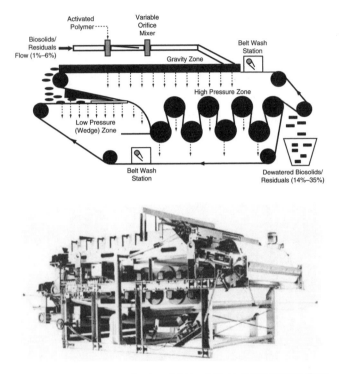

Courtesy of Ashbrook Simon-Hartley, Houston, Texas.

Operational Diagram and Photograph of a Belt Filter Press With Two Continuous Belts for Gravity and Pressure Dewatering With Uniform-Diameter Rollers

Typical Data for Various Types of Sludges Dewatered on Belt Filter Presses

Type of Wastewater Sludge	Total Feed Solids, %	Polymer, g/kg	Total Cake Solids, %
Raw primary	3–10	1–5	28–44
Raw waste-activated solids (WAS)	0.5–4	1–10	20–35
Raw primary + WAS	3–6	1–10	20–35
Anaerobically digested primary	3–10	1–5	25–36
Anaerobically digested WAS	3–4	2–10	12–22
Anaerobically digested primary + WAS	3–9	2–8	18–44
Aerobically digested primary + WAS	1–3	2–8	12–20
Oxygen-activated WAS	1–3	4–10	15–23
Thermally conditioned primary + WAS	4–8	0	25–50

Typical Operating Parameters for Belt Filter Press Dewatering of Polymer Flocculated Wastewater Sludges

Type of Sludge	Feed Solids, %	Hydraulic loading, gpm/m*	Solids Loading, lb/m·hr†	Cake Solids, %	Polymer Dosage, lb/ton‡
Anaerobically digested primary only	4–6	40–60	1,000–1,600	20–30	3–8
Anaerobically digested primary plus waste activated	2–5	40–60	500–1,000	15–26	8–14
Aerobically digested without primary	1–3	30–45	200–500	12–18	8–14
Raw primary and waste activated	3–6	40–50	800–1,200	18–26	4–10
Thickened waste activated	3–5	40–50	800–1,000	14–20	6–8
Extended aeration waste activated	1–3	30–50	200–600	12–22	8–14

Courtesy of Pearson Education, Inc.

* 1.0 gpm/m = 0.225 m³/m·hr
† 1.0 lb/m/hr = 0.454 kg/m·hr
‡ 1.0 lb/ton = 0.500 kg/tonne

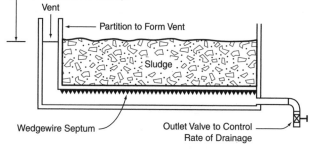

Cross-Section of a Wedgewire Drying Bed

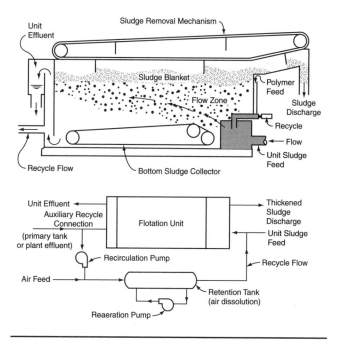

Dissolved Air Flotation Thickener

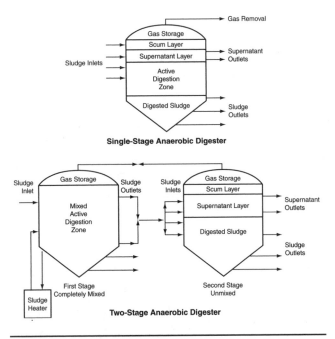

Single-Stage Anaerobic Digester

Two-Stage Anaerobic Digester

Configuration of Anaerobic Digesters

Anaerobic Lagoons

Some advantages and disadvantages of anaerobic lagoons are listed in the following sections.

Advantages

- More effective for rapid stabilization of strong organic wastes, making higher influent organic loading possible
- Produce methane, which can be used to heat buildings, run engines, or generate electricity, but methane collection increases operational problems
- Produce less biomass per unit of organic material processed. Less biomass produced equates to savings in sludge handling and disposal costs.

- Do not require additional energy, because they are not aerated, heated, or mixed
- Less expensive to construct and operate
- Ponds can be operated in series.

Disadvantage

They require a relatively large area of land.

CENTRIFUGES

Range of Expected Centrifuge Performance

Type of Wastewater Solids	Feed, % total solids	Polymer, lb/dry ton of solids	Cake, % total solids
Primary undigested	4–8	5–30	25–40
Waste-activated solids (WAS) undigested	1–4	15–30	16–25
Primary + WAS undigested	2–4	5–16	25–35
Primary + WAS aerobic digested	1.5–3	15–30	16–25
Primary + WAS anaerobic digested	2–4	15–30	22–32
Primary anaerobic digested	2–4	8–12	25–35
WAS aerobic digested	1–4	20	18–21
High-temperature aerobic	4–6	20–40	20–25
High-temperature anaerobic	3–6	20–30	22–28
Lime stabilized	4–6	15–25	20–28

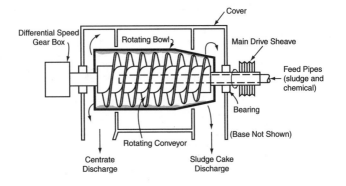

Solid Bowl Scroll Centrifuge

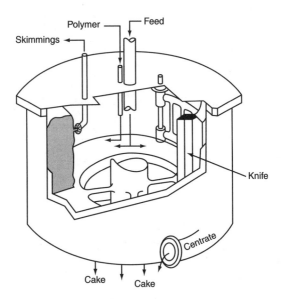

Imperforate Basket Centrifuge

MANAGEMENT PRACTICES

Storage

Field Storage (Stockpile) Checklist (involving dewatered cake, dried, or composted class A or class B biosolids)

Management	✓
1. Prepare and maintain a field management plan.	
2. Train employees to properly operate the site according to plan; conduct spill drills.	
3. Critical Control Point 1: Work with wastewater treatment plant to maximize biosolids stability, consistency, and quality; direct batches to appropriate sites.	
4. Critical Control Point 2: Transportation; clearly mark site access routes and stockpile areas; conduct spill drills.	
5. Maintain accurate and well-organized records.	
6. Designate a competent public relations person; maintain communication with stakeholders; notify agencies of reportable incidents; explain actions taken to respond to citizens' concerns or complaints.	
Operations	✓
1. Use biosolids that stay consolidated and nonflowing; shape stockpiles whenever possible to shed water.	
2. Minimize ponding and storage time to the extent feasible during hot, humid weather; manage accumulated water appropriately.	
3. Inspect and maintain upslope water diversions.	
4. Inspect buffer zones to ensure runoff is not moving out of bounds.	
5. Restrict public access and use temporary fencing to exclude livestock, where applicable; install signs; secure site appropriately.	
6. Clean all vehicles and equipment before they exit onto public roads.	
7. Train employees to use appropriate sanitation practices; inspect for use.	
8. Inspect for odors and conditions conducive to odors; apply chemicals or surface covering material to suppress odors if needed; consider the meteorological conditions and the potential for off-site odors when scheduling opening the storage pile and spreading of biosolids.	

Key Design Concepts for Constructed Biosolids Storage Facilities

Issue	Liquid/Thickened, 1%–12% solids		Dewatered/Dry Biosolids Facilities, 12%–30% solids/>50% solids (dry)		Liquid/Thickened, 1%–12% solids
	Lagoons	Pads/Basins	Enclosed Buildings	Tanks	
Design	Below-ground excavation. Impermeable liner of concrete, geotextile, or compacted earth.	Above ground. Impermeable liner of concrete, asphalt, or compacted earth.	Roofed, open-sided, or enclosed. Flooring: concrete, asphalt, or compacted earth.		Above or below ground. Concrete, metal or prefabricated. If enclosed, ventilation needed.
Capacity	Expected biosolids volume plus expected precipitation plus freeboard	Expected biosolids volume, unless precipitation is retained; then, biosolids volume plus expected precipitation plus freeboard	Expected biosolids volume		Enclosed: expected biosolids volume. If open top, expected biosolids volume plus expected precipitation plus freeboard.
Accumulated water management	Pump out and spray-irrigate or land-apply the liquid, haul to wastewater treatment plant (WWTP), or mix with biosolids	Sumps/pumps if facility is a basin for collection of water for spray irrigation; land-apply or haul to WWTP	Roof and gutter system, enclosure, or upslope diversions		Decant and spray-irrigate, land-apply, or haul to WWTP, or mix with biosolids in tank.

Table continued on next page

Key Design Concepts for Constructed Biosolids Storage Facilities (continued)

| Issue | Liquid/Thickened, 1%–12% solids | | Dewatered/Dry Biosolids Facilities, 12%–30% solids/>50% solids (dry) | | Liquid/Thickened, 1%–12% solids |
	Lagoons	Pads/Basins	Enclosed Buildings	Tanks
Runoff management	Diversions to keep runoff out of lagoon	Diversions to keep runoff out of sight; curbs and/or sumps to collect water for removal or downslope filter strips or treatment ponds	Enclosure or upslope diversions		Prevent gravity outflows from pipes and fittings. Diversions for open, below-ground tanks.
Biosolids consistency	Liquid or dewatered. Removal with pumps, cranes, or loaders.	If no side walls, material must stack without flowing.	Material must stack well enough to remain inside.		Liquid or dewatered biosolids. If enclosed, material must be liquid enough to pump.
Safety	Drowning hazard. Post warnings; fence; locked gates and rescue equipment on site.	Drowning hazard. Post warnings; fence; locked gates and rescue equipment on site.	Post "no trespassing" signs; remote location; lock doors, gates, and fences.		Posted warning; locking access points (e.g., use hatches, controlled access ladders, and confined-space entry procedures to access).

Biosolids

319

Constructed Facilities Checklist (involving lagoons, pads, or storage tanks)

Project Management	✓
1. Prepare and maintain a storage site management plan with spill plan.	
2. Critical Control Point 1: Work closely with the wastewater treatment plant on stability and consistency.	
3. Critical Control Point 2: Transportation—clearly mark site access routes and unloading areas.	
4. Train employees to properly operate the storage facility and to perform inspections; conduct spill drills.	
5. Maintain accurate and well-organized records.	
6. Designate a competent public relations person; maintain communications with stakeholders; notify agencies of reportable incidents; explain actions taken to respond to citizens' concerns or complaints.	
Operations	✓
1. Minimize ponding and storage time; manage accumulated water properly.	
2. Inspect and maintain up- and downslope water diversion/collection systems.	
3. Inspect and maintain tanks, ponds, curbs, gutters, and sumps used to collect runoff.	
4. Inspect buffer zones to ensure flow is not moving out of bounds.	
5. Install signs and implement security measures to restrict public access.	
6. Inspect concrete, wood, earth, walls, foundation, and monitoring wells at constructed storage facilities.	
7. Meet nutrient and hydraulic loading limits and state/local requirements when land-applying accumulated water from storage.	
8. Clean all vehicles and equipment before they exit onto public roads.	
9. Train employees to use appropriate sanitation practices; ensure practices are properly followed.	
10. If the characteristics of the biosolids have changed significantly during storage, retest nutrient and solids content prior to land application to recalculate land-application rate of biosolids.	
11. Inspect for odors and conditions conducive to odors; mitigate appropriately.	
12. Attend to site aesthetics.	

Practices to Prevent Mud or Biosolids From Being Tracked Onto Public Roadways

- Vehicles transporting biosolids should be cleaned before they leave the wastewater treatment plant.

- Concrete or asphalt off-loading pads at the storage facility will help keep equipment clean and make cleanup of drips or spills easier.

- The storage facility should have provisions to clean trucks and equipment when the need arises. Mud on tires or vehicles can be hand-scraped or removed with a high-pressure washer or with compressed air (as long as this does not exacerbate an existing dust problem).

- All vehicles should be inspected for cleanliness before leaving the site.

- Use mud flaps on the back of dump trailers to preclude biosolids getting on tires or undercarriage during unloading operations.

- Install a temporary gravel access pad as necessary at the entrance/exit to avoid soil ruts and tracking of mud onto roads.

- Public roadways accessing the site should be inspected each day during operational periods and cleaned promptly (shovel and sweep).

Minimizing Odor During Storage

- Stabilize biosolids at wastewater treatment plant as much as possible.

- Avoid use of polymers that lead to malodor.

- Maintain proper pH during treatment.

- Meet the vector attraction reduction requirements of the USEPA Part 503 Biosolids Rule.

- Locate storage at remote sites.

- Minimize duration of storage.

- Assess meteorological conditions before loading and unloading.

- Ensure good housekeeping.

Prevention and Management of Odorous Emissions Associated With Biosolids Stabilization or Processing Methods

Stabilization and Processing Methods	Potential Causes of Odorous Emissions	Long-Term Potential Solutions	Short-Term Temporary Solutions
Anaerobic digestion	"Sour," overloaded, or thermophilic digester; volatilization of fatty acids and sulfur compounds	Optimize digester; don't overload.	Apply topical lime to stored biosolids.
Aerobic digestion	Low solids retention time; high organic loading; poor aeration	Increase retention time and aeration; lower organic load.	
Drying beds	Incomplete digestion of biosolids being dried	Optimize digestion.	
Compost	Poor mixing of bulking agent; poor aeration; improperly operating biofilters	Mix better; adjust mix ratio and aeration rate; improve biofilter function.	Aerate more effectively; remix; recompost.
Alkaline stabilization	Addition of insufficient alkaline material so pH drops below 9, microbial decomposition may occur with generation of odorous compounds. Check compatibility of polymer with high pH and other additives (e.g., $FeCl_3$).	Increase pH. Provide finer mesh grade of alkaline material and mix better to avoid inadequate contact with biosolids.	Check pH; apply topical lime.
Thermal conditioning and drying	High-temperature volatilization of fatty acids and sulfur compounds	Use secondary treatment biosolids; primary solids are less stable and more odorous when heated.	Apply topical lime if biosolids are still liquid.

Other Important Factors at the Wastewater Treatment Plant That Affect the Odor Potential of Biosolids

- Periodic changes in influent characteristics (e.g., fish wastes, textile wastes, and other wastewaters with high-odor characteristics).

- Type of polymer used and its susceptibility to decomposition and release of intense and pervasive odorants such as amines when biosolids are heated or treated with strong alkaline materials.

- Blending of primary and secondary biosolids that may create anaerobic conditions or stimulate a resumption of microbial decomposition.

- Completeness of blending and mixing, and quality of products used for stabilization (i.e., type of lime and granule size).

- Effectiveness and consistency of vector attraction reduction (VAR) method, use of USEPA Part 503 Biosolids Rule VAR options 1–8 (treatment at wastewater treatment plant [WWTP]) versus VAR options 9–10 (at land application site).

- Handling, storage time, and storage method when stabilized biosolids are held at the WWTP prior to transport (e.g., anaerobic conditions developing in enclosed holding tanks when material is held for several days during hot weather).

Biosolids

Practices to Reduce the Potential for Unacceptable Off-Site Odors

- Ensure that the wastewater treatment plant has used processes that minimize odor during processing.

- Minimize storage time.

- Monitor and manage any water to prevent stagnant septic water accumulations.

- Avoid or minimize storage of biosolids during periods of hot and humid weather, if possible. During warm weather, check for odors frequently. Use lime or other materials to control odors before they reach unacceptable levels off-site.

- Empty constructed storage facilities as soon as possible in the spring for cleaning and inspection; keep idle until the following winter, if possible.

- Select remote sites with generous buffers between sensitive-neighbor areas.

- Consider weather conditions, prevailing wind directions, and the potential for off-site odors when scheduling and conducting cleanout/spreading operations. For example, operations on a hot, humid day, with an air inversion layer and wind moving in the direction of a residential area on the day of the block party, greatly increases the risk of odor complaints.

- Conduct loading/unloading and spreading operations as quickly and efficiently as possible to minimize the time that odors may be emitted. Surface crusts on stored biosolids seal in odors, but they break during handling and odors can be released.

- Enclosed handling or pumping systems at constructed facilities may reduce the potential for odors on a day-to-day basis, but these facilities still have the potential for odors during off-loading operations when active ventilation is used.

- Observe good housekeeping practices during facility loading and unloading. Clean trucks and equipment regularly to prevent biosolids buildup that may give rise to odors. If biosolids spills occur, clean them up promptly.

- Provide local government and state agency representatives with a contact name and number. Ask them to call the storage facility operator immediately if they receive citizen questions, concerns, or odor complaints resulting from storage of biosolids. Operator staff should politely receive citizen questions or complaints, collect the individual's name and phone number, conduct a prompt investigation, undertake control measures, if necessary, follow up with the person who filed the complaint, and document the event and actions.

Odor Remediation Measures for Use During Handling Operations

- Immediately correct any poor housekeeping problems (such as dirty equipment).

- Immediately treat any accumulated water that has turned septic with lime, chlorine, potassium permanganate, or other odor-control product; remove the water as quickly as possible to a suitable land application site.

- If odors are arising from lime-stabilized biosolids, pH should be measured. If it has dropped below 9.0, lime can be topically applied to dewatered material, or, in highly liquid systems, lime slurry can be blended into the biosolids by circulation. The pH should be monitored and dosed with lime until the desired pH has been achieved. Raising pH halts organic matter decomposition in the biosolids that can generate odorous compounds.

- For most types of biosolids (digested, lime stabilized, liquid, dewatered), applying a topical lime slurry will raise surface pH levels, create a crust, and reduce odors. Topical spray applications of potassium permanganate ($KMnO_4$) or enzymatic odor control products to neutralize odorous compounds may also be effective in some situations.

- Cover biosolids with compost or sawdust.

- If the odor is due to the combination of wind and weather conditions (hot, humid) and agitation and circulation of biosolids as part of unloading operations, it may be advisable to cease unloading operations until weather conditions are less likely to transport odors to sensitive off-site receptors.

- Spread and incorporate or inject odorous material as quickly as possible.

- For enclosed storage facilities, absorptive devices (charcoal or biofilters) incorporated into a ventilation system may be a feasible option for reducing odorous emissions.

- Cause the wastewater treatment plant to change its processes to produce less odorous biosolids.

Biosolids

Selected Odorous Compounds Observed in Association With Manure, Compost, Sewage Sludge, and Biosolids With Corresponding Ranges of Odor Threshold Values

		Odor Threshold
Compound	**Odor Character**	*µL/L (µg/L)*
Nitrogenous compounds		
Ammonia	Sharp pungent	5.2* (150)
Butylamine	Sour, ammonia-like	1.8* (6,200)
Dibutylamine	Fishy	(0.016)*
Diisopropylamine	Fishy	1.8* (1,300)
Dimethylamine	Putrid, fishy	0.13 (470)
Ethylamine	Ammonical	0.95* (4,300)
Methylamine	Putrid, fish	3.2* (2,400)
Triethylamine	Ammonical, fishy	0.48* (0.42)
Trimethylamine	Ammonical, fishy	0.00044*
Nitrogenous heterocyclics		
Indole	Fecal, nauseating	(0.00012–0.0015)†
Pyridine	Disagreeable, burnt, pungent	0.17* (0.95)
Skatole	Fecal, nauseating	(0.00035–0.0012)†
Sulfur-containing compounds		
Allyl mercaptan	Strong garlic, coffee	(0.000005)†
Amyl mercaptan	Unpleasant, putrid	(0.0003)†
Benzyl mercaptan	Unpleasant, strong	(0.013)†
Crotyl mercaptan	Skunk-like	(0.00000043)†
Dimethyl disulfide	Vegetable sulfide	(1.00)†
Dimethyl sulfide	Decayed vegetables	(0.0003–0.016)†
Diphenyl sulfide	Unpleasant	(0.0026)†
Ethyl mercaptan	Decayed cabbage	0.00076* (0.0000075)
Hydrogen sulfide	Rotten eggs	8.1* (0.000029)

Table continued on next page

Selected Odorous Compounds Observed in Association With Manure, Compost, Sewage Sludge, and Biosolids With Corresponding Ranges of Odor Threshold Values (continued)

Compound	Odor Character	Odor Threshold µL/L (µg/L)	
Methyl mercaptan	Decayed cabbage, sulfidy	0.0016*	(0.000024)
n-butyl mercaptan	Skunk, unpleasant	0.00097	(0.000012)
Propyl mercaptan	Unpleasant	0.0000025–0.000075	
Sulfur dioxide	Pungent, irritating	1.1*	(0.11)
Thiocresol	Skunk, rancid	(0.0001)[†]	
Thiophenol	Putrid, garlic-like	(0.00014)[†]	
Other chemicals or compounds			
Acetaldehyde	Pungent, fruity	0.050*	(0.034)
Chlorine	Pungent, suffocating	0.31*	(0.0020)
m-Cresol	Tar-like, pungent	0.000049–0.0079	(37)
n-butyl alcohol	Alcohol	0.84*	

*Microliters per liter is the odor threshold for dilutions in odor-free air, and micrograms per liter is the odor threshold; both units are equivalent to parts per million.

† Converted from weight-by-volume concentration (milligrams per cubi meter) to micrograms per liter.

Nutrient Content of Various Organic Materials

Material	N	P_2O_5	K_2O	Ca	Mg	S	Cl
Apple pomace	2	—	0.2	—	—	—	—
Blood (dried)	12–15	3.0	—	0.3	—	—	0.6
Bone meal (raw)	3.5	22.0	—	22.0	0.6	0.2	0.2
Bone meal (steamed)	2.0	28.0	0.2	23.0	0.3	0.1	—
Brewers grains (wet)	0.9	0.5	—	—	—	—	—
Common crab waste	2.0	3.6	0.2	—	—	—	—
Compost (garden)	Varies with feedstocks and processes						
Cotton waste from factory	1.3	0.4	0.4	—	—	—	—
Cottonseed meal	6–7	2.5	1.5	0.4	0.9	0.2	—
Cotton motes	2.0	0.5	3.0	4.0	0.7	0.6	—
Cowpea forage	0.4	0.1	0.4	—	—	—	—
Dog manure	2.0	10.0	0.3	—	—	—	—
Eggs	2.1	0.4	0.2	—	—	—	—
Egg shells	1.2	0.4	0.2	—	—	—	—

Table continued on next page

Nutrient Content of Various Organic Materials (continued)

Material	N	P_2O_5	K_2O	Ca	Mg	S	Cl
Feathers	15.0	—	—	—	—	—	—
Fermentation sludges	3.5	0.5	0.1	7.3	0.1	—	—
Fish scrap (dried)	9.5	6.0	—	6.1	0.3	0.2	1.5
Fly ash							
Coal	0.3	0.1	—	0.6	0.1	10.0	0.5
Wood	9.8	—	0.7	—	—	—	—
Frittercake							
Citric acid production	—	—	5.2	—	—	—	—
Enzyme production	—	—	2.2	2.0	0.5	0.3	—
Garbage tankage	2.5	1.5	1.0	3.2	0.3	0.4	1.3
Greensand	—	1–2	5.0	—	—	—	—
Hair	12–16	—	—	—	—	—	—
Legumes	3.0	1.5	1.0	0.5	2.4	1.9	1.2
Grass	0.8	0.2	0.2	0.3	0.2	—	—

Table continued on next page

Nutrient Content of Various Organic Materials (continued)

Material	Percentage by Weight						
	N	P₂0₅	K₂0	Ca	Mg	S	Cl
Oak leaves	0.8	0.4	0.2	—	—	—	—
Oyster shell siftings	0.4	10.4	0.1	—	—	—	—
Peanut hull meal	1.2	0.5	0.8	—	—	—	—
Peat/muck	2.7	—	—	0.7	0.3	1.0	0.1
Pine needles	0.5	0.1	—	—	—	—	—
Dissolved air flotation sludge	8.0	1.8	0.3	—	—	—	—
Potato tubers	0.4	0.2	0.5	—	—	—	—
Potato leaves and stalks	—	0.6	0.2	0.4	—	—	—
Potato skins, raw ash	—	—	5.2	2.0	7.5	—	—
Sawdust	0.2	—	0.2	—	—	—	—
Sea marsh hay	1.1	0.2	0.8	—	—	—	—
Seaweed (dried)	0.7	0.8	5.0	—	—	—	—
Sewage sludge (municipal)	2.6	3.7	0.2	1.3	0.2	—	—
Shrimp waste	2.9	10.0	—	—	—	—	—

Table continued on next page

Nutrient Content of Various Organic Materials (continued)

Material	Percentage by Weight						
	N	P_2O_5	K_2O	Ca	Mg	S	Cl
Soot from chimney	—	0.5–11	—	1.0	0.4	—	—
Soybean meal	7.0	1.2	1.5	0.4	0.3	0.2	—
Spent brewery yeast	—	7.0	0.4	0.3	0.04	0.03	—
Sweet potatoes	0.2	0.1	0.5	—	—	—	—
Tankage	7.0	1.5	3–10	—	—	—	—
Textile sludge	2.8	2.1	0.2	0.5	0.2	—	—
Wood ashes	0.0	2.0	6.0	20.0	1.0	—	—
Wood processing wastes	—	0.4	0.2	0.1	1.1	0.2	—
Tobacco stalks, leaves	3.7–4.0	0.5–0.6	4.5–6.0	—	—	—	—
Tobacco stems	2.5	0.9	7.0	—	—	—	—
Tomatoes, fruit, leaves	0.2–0.4	0.1	0.4	—	—	—	—

NOTE: Approximate values are given. Have materials analyzed for nutrient content before using.

Major Pathogens Potentially Present in Municipal Wastewater and Manure[*]

Pathogen	Disease/Symptoms for Organism
Bacteria	
Salmonella spp.	Salmonellosis (food poisoning), typhoid
Shigella spp.	Bacillary dysentery
Yersinia spp.	Acute gastroenteritis (diarrhea, abdominal pain)
Vibrio cholerae	Cholera
Campylobacter jejuni	Gastroenteritis
Escherichia coli (enteropathogenic)	Gastroenteritis
Viruses	
Poliovirus	Poliomyelitis
Coxsackievirus	Meningitis, pneumonia, hepatitis, fever, etc.
Echovirus	Meningitis, paralysis, encephalitis, fever, etc.
Hepatitis A virus	Infectious hepatitis
Rotavirus	Acute gastroenteritis with severe diarrhea
Norwalk agents	Epidemic gastroenteritis with severe diarrhea
Reovirus	Respiratory infections, gastroenteritis
Protozoa	
Cryptosporidium	Gastroenteritis
Entamoeba histolytica	Acute enteritis
Giardia lamblia	Giardiasis (diarrhea and abdominal cramps)
Balantidium coli	Diarrhea and dysentery
Toxoplasma gondii	Toxoplasmosis
Helminth Worms	
Ascaris lumbricoides	Digestive disturbances, abdominal pain
Ascaris suum	Can have symptoms: coughing, chest pain
Trichuris trichiura	Abdominal pain, diarrhea, anemia, weight loss
Toxocara canis	Fever, abdominal discomfort, muscle aches
Taenia saginata	Nervousness, insomnia, anorexia
Taenia solium	Nervousness, insomnia, anorexia
Necator americanus	Hookworm disease
Hymenolepis nana	Taeniasis

[*] Not all pathogens are necessarily present in all biosolids and manures, all the time.

Composting Basics

During composting, microorganisms break down organic matter in wastewater solids into carbon dioxide, water, heat, and compost. To ensure optimal conditions for microbial growth, carbon and nitrogen must be present in the proper balance in the mixture being composted. The ideal carbon-to-nitrogen ratio ranges from 25 to 35 parts carbon for each 1 part of nitrogen by weight. A lower ratio can result in ammonia odors. A higher ratio will not create optimal conditions for microbial growth causing degradation to occur at a slower rate and temperatures to remain below levels required for pathogen destruction. Wastewater solids are primarily a source of nitrogen and must be mixed with a higher carbon-containing material such as wood chips, sawdust, newspaper, or hulls. In addition to supplying carbon to the composting process, the bulking agent serves to increase the porosity of the mixture. Porosity is important to ensure that adequate oxygen reaches the composting mass. Oxygen can be supplied to the composting mass through active means such as blowers and piping or through passive means such as turning to allow more air into the mass. The proper amount of air along with biosolids and bulking agent is important.

Comparison of Composting Methods

Aerated Static Pile	Windrow	In-Vessel
Highly affected by weather (can be lessened by covering, but at increased cost)	Highly affected by weather (can be lessened by covering, but at increased cost)	Only slightly affected by weather
Extensive operating history, both small and large scale	Proven technology on small scale	Relatively short operating history compared to other methods
Large volume of bulking agent required, leading to large volume of material to handle at each stage (including final distribution)	Large volume of bulking agent required, leading to large volume of material to handle at each stage (including final distribution)	High biosolids-to-bulking-agent ratio, so less volume of material to handle at each stage
Adaptable to changes in biosolids and bulking agent characteristics	Adaptable to changes in biosolids and bulking agent characteristics	Sensitive to changes in characteristics of biosolids and bulking agents
Wide-ranging capital cost	Low capital costs	High capital costs
Moderate labor requirements	Labor intensive	Not labor intensive
Large land area required	Large land area required	Small land area adequate
Large volumes of air to be treated for odor control	High potential for odor generation during turning; difficult to capture/contain air for treatment	Small volume of process air that is more easily captured for treatment
Moderately dependent on mechanical equipment	Minimally dependent on mechanical equipment	Highly dependent on mechanical equipment
Moderate energy requirement	Low energy requirements	Moderate energy requirement

Methods for Meeting 40 CFR 504 Pathogen Requirements

The USEPA 40 CFR 503 regulations, specifically 503.32(a) and (b), require biosolids intended for agricultural use to meet certain pathogen and vector attraction reduction conditions. The intent of a Class A pathogen requirement is to reduce the level of pathogenic organisms in the biosolids to *below detectable levels*. The intent of the Class B requirements is to ensure that pathogens have been reduced to levels that are unlikely to pose a threat to public health and the environment under the specific use conditions. For Class B material that is land applied, site restrictions are imposed to minimize the potential for human and animal contact with the biosolids for a period of time following land application until environmental factors have further reduced pathogens. No site restrictions are required with Class A biosolids. Class B biosolids cannot be sold or given away in bags or other containers. The criteria for meeting Class A requirements are shown in the table on page 336, and the criteria for Class B are shown in the table on page 336.

Biosolids

Criteria for Meeting Class A Requirements

Parameter	Unit	Limit
Fecal coliform or *Salmonella*	MPN/g TS*	1,000
	MPN/4 g TS	3
AND one of the following process options:		
Temperature/time based on % solids	Alkaline treatment	
Prior test for enteric virus/viable helminth	Post-test for enteric virus/viable helminth ova	
Composting	Heat drying	
Heat treatment	Thermophilic aerobic digestion	
Beta ray irradiation	Gamma ray irradiation	
Pasteurization	Process to further reduce pathogens equivalent process	

* Most probable number per gram dry weight of total solids.

Criteria for Meeting Class B Requirements

Parameter	Unit	Limit
Fecal coliform	MPN or cfu/g TS*	2,000,000
OR one of the following process options:		
Aerobic digestion	Air drying	
Anaerobic digestion	Composting	
Lime stabilization	Process to significantly reduce pathogens equivalent	

* Most probable number or colony-forming units per gram dry weight of total solids.

Summary of Requirements for Vector Attraction Reduction Options

Option	Requirement	Where/When Requirements Must Be Met
Volatile Solids (VS) Reduction	≥38% VS reduction during solids treatment	Across the process
Anaerobic bench-scale test	<17% VS loss, 40 days at 30°–37°C (86°–99°F)	On anaerobic digested biosolids
Aerobic bench-scale test	<15% VS reduction, 30 days at 20°C (68°F)	On aerobic digested biosolids
Specific oxygen uptake rate (SOUR)	SOUR at 20°C (68°F) is ≤1.5 mg oxygen/hr/g total solids	On aerobic stabilized biosolids
Aerobic process	≥14 days at >40°C (104°F) with an average >45°C (113°F)	On composted biosolids
pH adjustment	≥12 measured at 25°C (77°F),* and remain at pH >12 for 2 hours and ≥11.5 for 22 more hours	When produced or bagged
Drying without primary solids	≥75% Total Solids (TS) prior to mixing	When produced or bagged
Drying with primary solids	≥90% TS prior to mixing	When produced or bagged

Table continued on next page

Biosolids

Summary of Requirements for Vector Attraction Reduction Options (continued)

Option	Requirement	Where/When Requirements Must Be Met
Soil injection	No significant amount of solids is present on the land surface 1 hour after injection. Class A biosolids must be injected within 8 hours after the pathogen reduction process.	When applied
Soil incorporation	≤6 hours after land application; Class A biosolids must be applied on the land within 8 hours after being discharged from the treatment process.	After application
Daily cover at field site	Biosolids placed on a surface disposal site must be covered with soil or other material at the end of each operating day.	After placement
pH adjustment of septage	≥12 measured at 25°C (77°F),* and remain at ≥12 for 30 minutes without addition of more alkaline material.	Septage

* Or corrected to 25°C.

Ideal Operating Ranges for Methane Fermentation

Parameter	Optimum	Extreme
Temperature, °C	30–35	25–40
pH	6.6–7.6	6.2–8.0
Alkalinity, mg/L as CaCO₃	2,000–3,000	1,000–5,000
Volatile acids, mg/L as acetic acid	50–500	2,000

Performance for Various Types of Domestic Wastewater Solids

Type of Wastewater Solids	Feed Total Solids, %	Typical Cycle Time, hr	Cake Total Solids, %
Primary + waste-activated solids (WAS)	3–8	2–2.5	45–50
Primary + WAS + trickling filter	6–8	1.5–3	35–50
Primary +WAS + FeCl₃	5–8	3–4	40–45
Primary +WAS + FeCl₃ – digested	6–8	3	40
Tertiary with lime	8	1.5	55
Tertiary with aluminum	4–6	6	36

Biosolids

Typical Design Criteria for Class B Alkaline Stabilization

Parameter	Design Criterion
Alkaline dose	0.25 lb/lb of wastewater solids at 20% solids
Retention time in mixer	1 minute
Retention time in curing vessel	30 minutes

Typical Biosolids Application Scenarios

Type of Site/ Vegetation	Schedule	Application Frequency	Application Rate
Agricultural land			
Corn	April, May, after harvest	Annually	5–10 dry tons/acre
Small grains	March–June, August, fall	Up to 3 times per year	2–5 dry tons/acre
Soybeans	April–June, fall	Annually	5–20 dry tons/acre
Hay	After each cutting	Up to 3 times per year	2–5 dry tons/acre
Forest land	Year round	Once every 2–5 years	5–100 dry tons/acre
Range land	Year round	Once every 1–2 years	2–60 dry tons/acre
Reclamation sites	Year round	Once	60–100 dry tons/acre

General Requirements for Land Application of Sewage Sludge

(a) No person shall apply sewage sludge to the land except in accordance with USEPA Part 503 Biosolids Rule land application requirements.

(b) No person shall apply bulk sewage sludge that is nonexceptional quality for pollutants (i.e., subject to cumulative pollutant loading rates in 503.13(b)(2)) to agricultural land, forest, a public contact site, or a reclamation site if any of the cumulative pollutant loading rates in 503.13(b)(2) have been reached.

(c) No person shall apply domestic septage to agricultural land, forest, or a reclamation site during a 365-day period if the annual application rate in 503.13(c) has been reached during that period.

(d) The person who prepares bulk sewage sludge that is applied to agricultural land, forests, areas where the potential for contact with the public is high (i.e., public contact site), or a reclamation site shall provide the person who applies the bulk sewage sludge written notification of the concentration of total nitrogen (as N on a dry weight basis) in the bulk sewage sludge.

(e)(1) The person who applies sewage sludge to the land shall obtain information needed to comply with applicable Part 503 requirements.

(e)(2)(i) Before bulk sewage sludge that is subject to cumulative pollutant loading rates (CPLRs) in 503.13(b)(2) is applied to the land, the person who proposes to apply the bulk sewage sludge shall contact the permitting authority for the state in which the bulk sewage sludge is being applied, to determine whether bulk sewage sludge subject to the cumulative pollutant loading rates in 503.13(b)(2) has been applied to the site since July 20, 1993.

(ii) If bulk sewage sludge subject to CPLRs has not been applied to the site since July 20, 1993, the cumulative amount of each pollutant listed in Table 2 of 503.13 may be applied to the site in accordance with 503.13(a)(2)(i).

(iii) If bulk sewage sludge subject to CPLRs in 503.13(b)(2) has been applied to the site since July 20, 1993, and the cumulative amount of each pollutant applied to the site since that date is known, the cumulative amount of each pollutant applied to the site shall be used to determine the additional amount of each pollutant that can be applied to the site in accordance with 503.13(a)(2)(i).

(iv) If bulk sewage sludge subject to CPLRs in 503.13(b)(2) has been applied to the site since July 20, 1993, and the cumulative amount of each pollutant applied to the site since that date is not known, sewage sludge subject to CPLRs may no longer be applied to the site.

Table continued on next page

(f) A person who prepares bulk sewage sludge shall provide the person who applies the bulk sewage sludge notice and necessary information to comply with applicable Part 503 requirements.

(g) When the person who prepares sewage sludge gives the material to another person who prepares sewage sludge, the person who provides the sewage sludge shall provide to the person who receives sewage sludge notice and necessary information to comply with the applicable Part 503 requirements.

(h) The person who applies bulk sewage sludge to the land shall provide the owner/leaseholder of the land on which the bulk sewage sludge is applied notice and necessary information to comply with applicable Part 503 requirements.

(i) Any person who prepares bulk sewage sludge that is applied to land in a state other than the state in which the bulk sewage sludge is prepared, shall provide written notice, prior to the initial application of bulk sewage sludge to the land application site by the applier, to the permitting authority for the state in which the bulk sewage sludge is proposed to be applied. The notice must include

(1) The location by either street address or latitude/longitude of each land application site.

(2) The approximate time period in which the bulk sewage sludge will be applied to the site.

(3) The name, address, telephone number, and National Pollutant Discharge Elimination System (NPDES) permit number (if appropriate) for the person who prepares the bulk sewage sludge.

(4) The name, address, telephone number, and NPDES permit number (if appropriate) for the person who will apply the bulk sewage sludge.

(j) Any person who applies bulk sewage sludge subject to the CPLRs in 503.13(b)(2) to the land shall provide written notice, prior to the initial application of the bulk sewage sludge to the application site by the applier, to the permitting authority for the state in which the bulk sewage sludge will be applied, and the permitting authority shall retain and provide access to the notice. The notice must include

(1) The location, by either street address or latitude/longitude, of each land application site.

(2) The name, address, and NPDES permit number (if appropriate) of the person who will apply the bulk sewage sludge.

Pollutant Limits for the Land Application of Sewage Sludge

	Concentration Limits	
Pollutant	**Ceiling Concentrations (Table 1 of 40 CFR 503.13), mg/kg (dry weight)**	**Pollutant Concentrations (PC) (Table 3 of 40 CFR 503.13) Monthly Average, mg/kg (dry weight)**
Arsenic	75	41
Cadmium	85	39
Chromium	3,000	1,200
Copper	4,300	1,500
Lead	840	300
Mercury	57	17
Molybdenum[*]	75	—
Nickel	420	420
Selenium	100	36
Zinc	7,500	2,800

	Loading Rates			
	Cumulative Pollutant Loading Rates (CPLRs) (Table 2 of 40 CFR 503.13)		**Annual Pollutant Loading Rates (RPLRs) (Table 4 of 40 CFR 503.13)**	
Pollutant	**kg/ha (dry weight)**	**lb/acre (dry weight)**	**kg/ha/365-day period (dry weight)**	**lb/acre/365-day period (dry weight)**
Arsenic	41	37	2.0	1.8
Cadmium	39	35	1.9	1.7
Chromium	3,000	2,677	150	134
Copper	1,500	1,339	75	67
Lead	300	268	15	13
Mercury	17	15	0.85	0.76
Molybdenum[*]	—	—	—	—
Nickel	420	375	21	19
Selenium	100	89	5.0	4.5
Zinc	2,800	2,500	140	125

[*] The pollutant concentration limit, cumulative pollutant loading rate, and annual pollutant loading rate for molybdenum were deleted from Part 503 effective February 19, 1994. USEPA will reconsider establishing these limits at a later date.

Exclusions From USEPA Part 503 Biosolids Rule

Part 503 Does Not Include Requirements for:	Applicable Federal Regulation
Treatment of Biosolids—Processes used to treat sewage sludge prior to final use or disposal (e.g., thickening, dewatering, storage, heat drying)	None (except for operational parameters used to meet the Part 503 pathogen and Vector attraction reduction requirements)
Selection of Use or Disposal Practice—The selection of a biosolids use or disposal practice	None (the determination of the biosolids use or the disposal practice is a local decision)
Incineration of Biosolids With Other Wastes—Biosolids co-fired in an incinerator with other wastes (other than as an auxiliary fuel)	40 CFR Parts 60, 61
Storage of Biosolids—Placement of biosolids on land for 2 years or less (or longer when demonstrated not to be a surface disposal site but rather, based on practices, constitutes treatment or temporary storage)	None
Industrial Sludge—Sludge generated at an industrial facility during the treatment of industrial wastewater with or without combined domestic sewage	40 CFR Part 257 if land applied: 40 CFR Part 258 if placed in a municipal solid waste landfill
Hazardous Sewage Sludge—Sewage sludge determined to be hazardous in accordance with 40 CFR Part 261, Identification and Listing of Hazardous Waste	40 CFR Parts 261–268

Table continued on next page

Exclusions From USEPA Part 503 Biosolids Rule (continued)

Part 503 Does Not Include Requirements for:	Applicable Federal Regulation
Sewage Sludge Containing PCBs ≥50 mg/kg—Sewage sludge with a concentration of polychlorinated biphenyls (PCBs) equal to or greater than 50 mg/kg of total solids (dry weight basis)	40 CFR Part 761
Incinerator Ash—Ash generated during the firing of biosolids in a biosolid incinerator	40 CFR Part 257 if land applied; 40 CFR Part 258 if placed in a municipal solid waste landfill or 40 CFR Parts 261–268 if hazardous
Grit and Screenings—Grit (e.g., sand, gravel, cinders) or screenings (e.g., relatively large materials such as rags) generated during preliminary treatment of domestic sewage in a treatment works	40 CFR Part 257 if land applied; 40 CFR Part 258 if placed in a municipal solid waste landfill
Drinking Water Sludge—Sludge generated during the treatment of either surface water or groundwater used for drinking water	40 CFR Part 257 if land applied; 40 CFR Part 258 if placed in a municipal solid waste landfill
Certain Nondomestic Septage—Septage that contains industrial or commercial septage, including grease-trap pumpings	40 CFR Part 257 if land applied; 40 CFR Part 258 if placed in a municipal solid waste landfill

Biosolids

Who Must Apply for a Permit?

Treatment works treating domestic sewage (TWTDS) *required* to apply for a permit

- All generators of biosolids that are regulated by USEPA Part 503 Biosolids Rule (including all publicly owned treatment works)

- Industrial facilities that *separately* treat domestic sewage and generate biosolids that are regulated by Part 503

- All surface disposal site owner/operators

- All biosolids incinerator owner/operators

- Any person (e.g., individual, corporation, or government entity) who changes the quality of biosolids regulated by Part 503 (e.g., biosolids blenders or processors)[*]

- Any other person or facility designated by the permitting authority as a TWTDS

TWTDS and other persons *not automatically required* to apply for a permit[†]

- Biosolids land appliers, haulers, persons who store, or transporters who do not generate or do not change the quality of the biosolids

- Landowners of property on which biosolids are applied

- Domestic septage pumpers/haulers/treaters/appliers

- Biosolids packagers/baggers (who do not change the quality of the biosolids)

[*] If all the biosolids received by a biosolids blender or composter are exceptional quality (EQ) biosolids, then no permit will be required for the person who receives or processes the EQ biosolids.

[†] USEPA may request permit applications from these facilities when necessary to protect public health and the environment from reasonably anticipated effects of pollutants that may be present in biosolids.

Types of Land Onto Which Different Types of Biosolids May Be Applied

Biosolids Option[*]	Pathogen Class	VAR[†] Options	Type of Land	Other Restrictions
EQ	A	1–8	All[‡]	None
PC	A	9 or 10	All except lawn and home gardens[§]	Management practices
	B	1–10	All except lawn and home gardens[§]	Management practices and site restrictions
CPLR	A	1–10	All except lawn and home garden[**]	Management practices
	B	1–10	All except lawn and home garden[§**]	Management practices and site restrictions
APLR	A	1–8	All, but most likely lawns and home gardens	Labeling management practice

[*] EQ = exceptional quality; PC = pollutant concentration; CPLR = cumulative pollutant loading rate; APLR = annual pollutant loading rate.

[†] VAR = vector attraction reduction.

[‡] Agricultural land, forest, reclamation sites, and lawns and home gardens.

[§] It is not possible to impose site restrictions on lawns and home gardens.

[**] It is not possible to track cumulative additions of pollutants on lawns and home gardens.

Biosolids

Solids Concentrations and Other Characteristics of Various Types of Sludge

Wastewater Treatment	Primary, Gravity	Secondary, Biological	Advanced (Tertiary), Chemical Precipitation, Filtration
Sludge			
Amounts generated, L/m^3 of wastewater	2.5–3.5	15–20	25–30
Solids content, %	3–7	0.5–2.0	0.2–1.5
Organic content, %	60–80	50–60	35–50
Treatability, relative	Easy	Difficult	Difficult
Dewatered by belt filter			
Feed solids, %	3–7	3–6	
Cake solids, %	28–44	20–35	

Options for Meeting Pollutant Limits and Pathogen and Vector Attraction Reduction Requirements for Land Application

Biosolids Option*	Pollutant Limits	Pathogen Requirements	Vector Attraction Reduction Requirements
EQ	Bulk or bagged biosolids meet PC limits in the table titled Pollutant Limits (page 343)	Any one of the class A requirements in the table titled Summary of Class A and Class B Pathogen Reduction Requirements (page 349)	Any one of options 1 through 8 in the table titled Summary of Vector Attraction Reduction Options (page 350)
PC	Bulk biosolids meet PC limits in the table titled Pollutant Limits (page 343)	Any one of the class B requirements in the tables titled Summary of Class A and Class B Pathogen Reduction Requirements (page 349) and Restrictions for the Harvesting of Crops and Turf (page 351)	Any one of the ten options in the table titled Summary of Vector Attraction Reduction Options (page 350)
		Any one of the class A requirements in the table titled Summary of Class A and Class B Pathogen Reduction Requirements (page 349)	Options 9 or 10 in the table titled Summary of Vector Attraction Reduction Options (page 350)
CPLR	Bulk biosolids applied subject to CPLR limits in the table titled Pollutant Limits (page 343)	Any one of the class A or class B requirements in the tables titled Summary of Class A and Class B Pathogen Reduction Requirements (page 349) and Restrictions for the Harvesting of Crops and Turf (page 351)	Any one of the ten options in the table titled Summary of Vector Attraction Reduction Options (page 350)
APLR	Bagged biosolids applied subject to APLR limits in the table titled Pollutant Limits (page 343)	Any one of the class A requirements in the table titled Summary of Class A and Class B Pathogen Reduction Requirements (page 349)	Any one of the first eight options in the table titled Summary of Vector Attraction Reduction Options (page 350)

* EQ = exceptional quality; PC = pollutant concentration; CPLR = cumulative pollutant loading rate; APLR = annual pollutant loading rate. Each of these options also requires that the biosolids meet the ceiling concentrations for pollutants listed in the table titled Pollutant Limits for the Land Application of Sewage Sludge, and that the frequency of monitoring requirements in the table titled Frequency of Monitoring for Pollutants, Pathogen Densities, and Vector Attraction Reduction and record-keeping and reporting requirements in the table titled Record-Keeping and Reporting Requirements be met. In addition, the general requirements in the table titled USEPA Part 503 Biosolids Rule Land Application General Requirements and the management practices in the table titled USEPA Part 503 Biosolids Rule Land Application Management Practice Requirements have to be met when biosolids are land-applied (except for EQ biosolids).

Summary of Class A and Class B Pathogen Reduction Requirements

Class A

In addition to meeting the requirements in one of the six alternatives listed below, fecal coliform or *Salmonella* sp. bacteria levels must meet specific density requirements at the time of biosolids use or disposal, or when prepared for sale or giveaway.

Alternative 1: Thermally Treated Biosolids

Use one of four time–temperature regimens.

Alternative 2: Biosolids Treated in a High pH–High Temperature Process

Specifies pH, temperature, and air-drying requirements.

Alternative 3: For Biosolids Treated in Other Processes

Demonstrate that the process can reduce enteric viruses and viable helminth ova. Maintain operating conditions used in the demonstration.

Alternative 4: Biosolids Treated in Unknown Processes

Demonstration of the process is unnecessary. Instead, test for pathogens— *Salmonella* sp. or fecal coliform bacteria, enteric viruses, and viable helminth ova—at the time the biosolids are used or disposed of, or are prepared for sale or giveaway.

Alternative 5: Use of Processes to Further Reduce Pathogens (PFRPs)

Biosolids are treated in one of the PFRPs (see the table titled Processes to Further Reduce Pathogens—page 362).

Alternative 6: Use of a Process Equivalent to PFRPs

Biosolids are treated in a process equivalent to one of the PFRPs, as determined by the permitting authority.

Class B

The requirements in one of the three alternatives below must be met.

Alternative 1: Monitoring of Indicator Organisms

Test for fecal coliform density as an indicator for all pathogens at the time of biosolids use or disposal.

Alternative 2: Use of Processes to Significantly Reduce Pathogens (PSRPs)

Biosolids are treated in one of the PSRPs (see the table titled Processes to Significantly Reduce Pathogens—page 364).

Alternative 3: Use of Processes Equivalent to PSRPs

Biosolids are treated in a process equivalent to one of the PSRPs, as determined by the permitting authority.

Summary of Vector Attraction Reduction Options

Requirements in one of the following options must be met:

Option 1: Reduce the mass of volatile solids by a minimum of 38%.

Option 2: Demonstrate vector attraction reduction (VAR) with additional anaerobic digestion in a bench-scale unit.

Option 3: Demonstrate VAR with additional aerobic digestion in a bench-scale unit.

Option 4: Meet a specific oxygen uptake rate for aerobically treated biosolids.

Option 5: Use aerobic processes at greater than 104°F (40°C) (average temperatures 113°F [45°C]) for 14 days or longer (e.g., during biosolids composting).

Option 6: Add alkaline materials to raise the pH under specified conditions.

Option 7: Reduce moisture content of biosolids that do not contain unstabilized solids from other than primary treatment to at least 75% solids.

Option 8: Reduce moisture content of biosolids with unstabilized solids to at least 90%.

Option 9: Inject biosolids beneath the soil surface within a specified time, depending on the level of pathogen treatment.

Option 10: Incorporate biosolids applied to or placed on the land surface within specified time periods after application to or placement on the land surface.

Restrictions for the Harvesting of Crops and Turf, Grazing of Animals, and Public Access on Sites Where Class B Biosolids Are Applied

Restrictions for the harvesting of crops and turf:

- Food crops, feed crops, and fiber crops, whose edible parts do not touch the surface of the soil, shall not be harvested until *30 days* after biosolids application.

- Food crops with harvested parts that touch the biosolids/soil mixture and are totally aboveground shall not be harvested until *14 months* after application of biosolids.

- Food crops with harvested parts below the land surface where biosolids remain on the land surface for 4 months or longer prior to incorporation into the soil shall not be harvested until *20 months* after biosolids application.

- Food crops with harvested parts below the land surface where biosolids remain on the land surface for less than 4 months prior to incorporation shall not be harvested until *38 months* after biosolids application.

- Turf grown on land where biosolids are applied shall not be harvested until *1 year* after application of the biosolids when the harvested turf is placed on either land with a high potential for public exposure or a lawn, unless otherwise specified by the permitting authority.

Restriction for the grazing of animals:

Animals shall not be grazed on land until *30 days* after application of biosolids to the land.

Restrictions for public contact:

- Access to land with a high potential for public exposure, such as a park or ballfield, is restricted for *1 year* after biosolids application. Examples of restricted access include posting with no-trespassing signs and fencing.

- Access to land with a low potential for public exposure (e.g., private farmland) is restricted for *30 days* after biosolids application. An example of restricted access is remoteness.

Biosolids

Examples of Crops Impacted by Site Restrictions for Class B Biosolids

Harvested Parts That		
Usually Do Not Touch the Soil/Biosolids Mixture	**Usually Touch the Soil/Biosolids Mixture**	**Are Below the Soil/Biosolids Mixture**
Peaches	Melons	Potatoes
Apples	Strawberries	Yams
Oranges	Eggplant	Sweet potatoes
Grapefruit	Squash	Rutabaga
Corn	Tomatoes	Peanuts
Wheat	Cucumbers	Onions
Oats	Celery	Leeks
Barley	Cabbage	Radishes
Cotton	Lettuce	Turnips
Soybeans		Beets

Procedure for the Applier to Determine the Amount of Nitrogen Provided by the AWSAR Relative to the Agronomic Rate

Assume that the annual whole sludge (biosolids) application rate (AWSAR) for biosolids is 410 lb of biosolids per 1,000 ft^2 of land. If biosolids were to be placed on a lawn that has a nitrogen requirement of about 200 lb[*] of available nitrogen per acre per year, the following steps would determine the amount of nitrogen provided by the AWSAR relative to the agronomic rate if the AWSAR was used:

1. The nitrogen content of the biosolids indicated on the label is 1% total nitrogen and 0.4% available nitrogen the first year.

2. The AWSAR is 410 lb of biosolids per 1,000 ft^2, which is 17,860 lb of biosolids per acre:

$$\frac{410 \text{ lb}}{1,000 \text{ sq ft}} \times \frac{43,560 \text{ sq ft}}{\text{acre}} \times 0.001 = \frac{17,860 \text{ lb}}{\text{acre}}$$

---

[*] Assumptions about crop nitrogen requirement, biosolids nitrogen content, and percent of that nitrogen that is available are for illustrative purposes only.

3. The available nitrogen from the biosolids is 71 lb/acre:

$$\frac{17{,}860 \text{ lb biosolids}}{\text{acre}} \times 0.004 = \frac{71 \text{ lb}}{\text{acre}}$$

4. Because the biosolids application will only provide 71 lb of the total 200 lb of nitrogen required, in this case the AWSAR for the biosolids will not cause the agronomic rate for nitrogen to be exceeded and an additional 129 lb/acre of nitrogen would be needed from some other source to supply the total nitrogen requirement of the lawn.

Frequency of Monitoring for Pollutants, Pathogen Densities, and Vector Attraction Reduction

Amounts of Biosolids, metric tons per 365-day period	Amount of Biosolids,* short tons		Frequency
	Average per Day	Per 365 Days	
>0 to <290	>0 to <0.85	>0 to <320	Once per year
≥290 to <1,500	0.85 to <4.5	320 to <1,650	Once per quarter (4 times per year)
≥1,500 to <15,000	4.5 to <45	1,650 to <16,500	Once per 60 days (6 times per year)
≥15,000	≥45	≥16,500	Once per month (12 times per year)

*Either the amount of bulk biosolids applied to the land or the amount of biosolids received by a person who prepares biosolids for sale or giveaway in a bag or other container for application to the land (dry weight basis).

For Exceptional Qualtiy (EQ) Biosolids

None (unless set by USEPA or state permitting authority on a case-by-case basis for bulk biosolids to protect public health and the environment).

For Pollutant Concentration (PC) and Cumulative Pollutant Loading Rate (CPLR) Biosolids

The preparer must notify and provide information necessary to comply with the Part 503 land application requirements to the person who applies bulk biosolids to the land.

The preparer* who provides biosolids to another person who further prepares the biosolids for application to the land must provide this person with notification and information necessary to comply with the Part 503 land application requirements.

The preparer must provide written notification of the total nitrogen concentration (as N on a dry weight basis) in bulk biosolids to the applier of the biosolids to agricultural land, forests, public contact sites, or reclamation sites.

The applier of biosolids must obtain information necessary to comply with the Part 503 land application requirements, apply biosolids to the land in accordance with the Part 503 land application requirements, and provide notice and necessary information to the owner or leaseholder of the land on which biosolids are applied.

Out-of-State Use

The preparer must provide written notification (prior to the initial application of the bulk biosolids by the applier) to the permitting authority in the state where biosolids are proposed to be land-applied when bulk biosolids are generated in one state and transferred to another state for application to the land. The notification must include all of the following:

- the location (either street address or latitude and longitude) of each land application site

- the approximate time period the bulk biosolids will be applied to the site

- the name, address, telephone number, and National Pollutant Discharge Elimination System (NPDES) permit number for both the preparer and the applier of the bulk biosolids

- additional information or permits in both states, if required by the permitting authority

Table continued on next page

Additional Requirements for CPLR Biosolids

The applier must notify the permitting authority in the state where bulk biosolids are to be applied prior to the initial application of the biosolids. This is a one-time notice requirement for each land application site each time there is a new applier. The notice must include each of the following:

- the location (either street address or latitude and longitude) of the land application site

- the name, address, telephone number, and NPDES permit number (if appropriate) of the person who will apply the bulk biosolids

The applier must obtain records (if available) from the previous applier, landowner, or permitting authority that indicate the amount of each CPLR pollutant in biosolids that have been applied to the site since July 20, 1993. In addition

- When these records are available, the applier must use this information to determine the additional amount of each pollutant that can be applied to the site in accordance with the CPLRs in the table titled Pollutant Limits (page 343).

- The applier must keep the previous records and also record the additional amount of each pollutant he or she is applying to the site.

- When records of past known CPLR applications since July 20, 1993, are not available, biosolids meeting CPLRs cannot be applied to that site. However, EQ or PC biosolids could be applied.

If biosolids meeting CPLRs have not been applied to the site in excess of the limit since July 20, 1993, the CPLR limit for each pollutant in the table titled Pollutant Limits (page 343) will determine the maximum amount of each pollutant that can be applied in biosolids if the following are true:

- all applicable management practices are followed

- the applier keeps a record of the amount of each pollutant in biosolids applied to any given site

The applier must not apply additional biosolids under the cumulative pollutant loading concept to a site where any of the CPLRs have been reached.

* The preparer is either the person who generates the biosolids or the person who derives a material from biosolids.

USEPA Part 503 Biosolids Rule Land Application Management Practice Requirements

For Exceptional Quality (EQ) Biosolids

None (unless established by USEPA or the state permitting authority on a case-by-case basis for bulk biosolids to protect public health and the environment).

For Pollutant Concentration and Cumulative Pollutant Loading Rate Biosolids

These types of biosolids cannot be applied to flooded, frozen, or snow-covered agricultural land, forests, public contact sites, or reclamation sites in such a way that the biosolids enter a wetland or other waters of the United States (as defined in 40 CFR Part 122.2, which generally includes tidal waters, interstate and intrastate waters, tributaries, the territorial sea, and wetlands adjacent to these waters), except as provided in a permit issued pursuant to Section 402 (National Pollutant Discharge Elimination System permit) or Section 404 (Dredge and Fill Permit) of the Clean Water Act, as amended.

These types of biosolids cannot be applied to agricultural land, forests, or reclamation sites that are 33 ft (10 m) or less from US waters, unless otherwise specified by the permitting authority.

If applied to agricultural lands, forests, or public contact sites, these types of biosolids must be applied at a rate that is equal to or less than the agronomic rate for nitrogen for the crop to be grown. Biosolids applied to reclamation sites may exceed the agronomic rate for nitrogen as specified by the permitting authority.

These types of biosolids must not harm or contribute to the harm of a threatened or endangered species or result in the destruction or adverse modification of the species' critical habitat when applied to the land. Threatened or endangered species and their critical habitats are listed in Section 4 of the Endangered Species Act. Critical habitat is defined as any place where a threatened or endangered species lives and grows during any stage of its life cycle. Any direct or indirect action (or the result of any direct or indirect action) in a critical habitat that diminishes the likelihood of survival and recovery of a listed species is considered destruction or adverse modification of a critical habitat.

For Annual Pollution Loading Rate (APLR) Biosolids

A label must be affixed to the bag or other container or an information sheet must be provided to the person who receives APLR biosolids in other containers. At a minimum, the label or information sheet must contain the following information:

- the name and address of the person who prepared the biosolids for sale or giveaway in a bag or other container
- a statement that prohibits application of the biosolids to the land except in accordance with the instructions on the label or information sheet
- an annual whole sludge (biosolids) application rate, or annual whole sludge application rate, for the biosolids that do not cause the APLRs to be exceeded
- the nitrogen content

There is no labeling requirement for EQ biosolids sold or given away in a bag or other container.

Record-Keeping and Reporting Requirements

Type of Biosolids	Records That Must Be Kept	Person Responsible for Record Keeping		Records That Must Be Reported[*]
		Preparer	Applier	
Exceptional quality (EQ)	Pollutant concentrations	X		X
	Pathogen reduction certification and description	X		X
	Vector attraction reduction (VAR) certification and description	X		X
Pollutant concentration (PC)	Pollutant concentrations	X		X
	Management practice certification and description		X	
	Site restriction certification and description (where class B pathogen requirements are met)		X	
	Pathogen reduction certification and description	X		X
	VAR certification and description	X	X[†]	X[‡]
Cumulative pollutant loading rate (CPLR)	Pollutant concentrations	X		X
	Management practice certification and description		X	
	Site restriction certification and description (if class B pathogen requirements are met)		X	
	Pathogen reduction certification and description	X		X
	VAR certification and description	X	X[†]	X[‡]

Table continued on next page

Biosolids

Record-Keeping and Reporting Requirements (continued)

Type of Biosolids	Records That Must Be Kept	Person Responsible for Record Keeping		Records That Must Be Reported*
		Preparer	Applier	
	Other information:			
	• Certification and description of information gathered (information from the previous applier, landowner, or permitting authority regarding the existing cumulative pollutant load at the site from previous biosolids applications)			
	• Site location			
	• Number of hectares		X	X§
	• Amount of biosolids applied			
	• Cumulative amount of pollutant applied (including previous amounts)			
	• Date of application			
Annual pollutant loading rate (APLR)	Pollutant concentrations	X		X
	Management practice certification and description	X		X
	Pathogen reduction certification and description	X		X
	VAR certification and description	X		X
	The annual whole sludge application rate for the biosolids	X		X

* Reporting responsibilities are only for publicly owned treatment works (POTWs) with a design flow rate ≥1 mgd, POTWs that serve a population of ≥10,000, and class 1 sludge management facilities.

† The preparer certifies and describes VAR methods other than injection and incorporation of biosolids into the soil. The applier certifies and describes injection or incorporation of biosolids into the soil.

‡ Records that certify and describe injection or incorporation of biosolids into the soil do not have to be reported.

§ Some of this information has to be reported only when 90% or more of any of the CPLRs is reached at a site.

358

Management Practices for Surface Disposal Sites

Biosolids placed on a disposal unit must not harm threatened or endangered species.

The active biosolids unit must not restrict base flood flow.

The active biosolids unit must be located in a geologically stable area:

- must not be located in an unstable area

- must not be located in a fault area with displacement in Holocene time (unless allowed by the permitting authority)

- if located in a seismic impact zone, must be able to withstand certain ground movements

The active biosolids unit cannot be located in wetlands (unless allowed in a permit).

Runoff must be collected from the surface disposal site with a system capability to handle a 25-year, 24-hour storm event.

Only where there is a liner, leachate must be collected and the owner/operator must maintain and operate a leachate collection system.

Only where there is a cover, there must be limits on concentrations of methane gas in air in any structure on the site and in air at the property line of the surface disposal site.

The owner/operator cannot grow crops on site (unless allowed by the permitting authority).

The owner/operator cannot graze animals on site (unless allowed by the permitting authority).

The owner/operator must restrict public access.

The biosolids placed in the active biosolids unit must not contaminate an aquifer.

Pathogen and Vector Attraction Reduction Requirements for Surface Disposal Sites

Pathogen Reduction Requirements

Options (must meet one of these):

- Place a daily cover on the active biosolids unit.

- Meet one of six class A pathogen reduction requirements (see the table titled Summary of Class A and Class B Pathogen Reduction Requirements—page 349)

- Meet one of three class B pathogen reduction requirements, except site restrictions (see the table titled Summary of the Three Alternatives for Meeting Class B Pathogen Requirements—page 362)

Vector Attraction Reduction Requirements

Options (must meet one of these):

- Place a daily cover on the active biosolids unit.

- Reduce volatile solids content by a minimum of 38% or less under specific laboratory test conditions with anaerobically and aerobically digested biosolids.

- Meet a specific oxygen uptake rate.

- Treat the biosolids in an aerobic process for a specified number of days at a specified temperature.

- Raise the pH of the biosolids with an alkaline material to a specified level for a specified time.

- Meet a minimum percent-solids content.

- Inject or incorporate the biosolids into soil.

The Four Time–Temperature Regimes for Class A Pathogen Reduction Under Alternative 1

Regime	Applies to:	Requirement	Time–Temperature Relationship[*]
A	Biosolids with 7% solids or greater (except those covered by regime B)	Temperature of biosolids must be 50°C or higher for 20 minutes or longer	$D = \dfrac{131,700,000}{10^{0.14t}}$
B	Biosolids with 7% solids or greater in the form of small particles and heated by contact with either warmed gases or an immiscible liquid	Temperature of biosolids must be 50°C or higher for 15 seconds or longer	$D = \dfrac{131,700,000}{10^{0.14t}}$
C	Biosolids with less than 7% solids	Heated for at least 15 seconds but less than 30 minutes	$D = \dfrac{131,700,000}{10^{0.14t}}$
D	Biosolids with less than 7% solids	Temperature of sludge is 50°C or higher with at least 30 minutes or longer contact time	$D = \dfrac{50,070,000}{10^{0.14t}}$

[*] D = time in days; t = temperature in degrees Celsius.

Biosolids

Processes to Further Reduce Pathogens Listed in Appendix B of 40 CFR Part 503

Composting

Using either the within-vessel composting method or the static aerated pile composting method, the temperature of the biosolids is maintained at 55°C or higher for 3 days.

Using the windrow composting method, the temperature of the biosolids is maintained at 55°C or higher for 15 days or longer. During the period when the compost is maintained at 55°C or higher, the windrow is turned a minimum of five times.

Heat Drying

Biosolids are dried by direct or indirect contact with hot gases to reduce the moisture content of the biosolids to 10% or lower. Either the temperature of the biosolids particles exceeds 80°C or the wet bulb temperature of the gas in contact with the biosolids as the biosolids leave the dryer exceeds 80°C.

Heat Treatment

Liquid biosolids are heated to a temperature of 180°C or higher for 30 minutes.

Thermophilic Aerobic Digestion

Liquid biosolids are agitated with air or oxygen to maintain aerobic conditions, and the mean cell residence time of the biosolids is 10 days at 55°–60°C.

Beta Ray Irradiation

Biosolids are irradiated with beta rays from an accelerator at dosages of at least 1.0 Mrad at room temperature (ca. 20°C).

Gamma Ray Irradiation

Biosolids are irradiated with gamma rays from certain isotopes, such as Cobalt 60 and Cesium 137, at room temperature (ca. 20°C).

Pasteurization

The temperature of the biosolids is maintained at 70°C or higher for 30 minutes or longer.

Summary of the Three Alternatives for Meeting Class B Pathogen Requirements

Alternative 1: The Monitoring of Indicator Organisms

Test for fecal coliform density as an indicator for all pathogens. The geometric mean of seven samples shall be less than 2 million most probable numbers per gram per total solids or less than 2 million colony-forming units per gram of total solids at the time of use or disposal.

Alternative 2: Biosolids Treated in Processes to Significantly Reduce Pathogens (PSRPs)

Biosolids must be treated in one of the PSRPs (see the table titled Process to Significantly Reduce Pathogens Listed in Appendix B of 40 CFR Part 503).

Alternative 3: Biosolids Treated in a Process Equivalent to a PSRP

Biosolids must be treated in a process equivalent to one of the PSRPs, as determined by the permitting authority.

Site Restrictions for Class B Biosolids Applied to the Land

Food Crops With Harvested Parts That Touch the Biosolids/Soil Mixture

Food crops with harvested parts that touch the biosolids/soil mixture and are totally above the land surface shall not be harvested for *14 months* after application of biosolids.

Food Crops With Harvested Parts Below the Land Surface

Food crops with harvested parts below the surface of the land shall not be harvested for *20 months* after application of the biosolids when the biosolids remain on the land surface for 4 months or longer prior to incorporation into the soil.

Food crops with harvested parts below the surface of the land shall not be harvested for *38 months* after application of biosolids when the biosolids remain on the land surface for less than 4 months prior to incorporation into the soil.

Food Crops With Harvested Parts That Do Not Touch the Biosolids/Soil Mixture, Feed Crops, and Fiber Crops

Food crops with harvested parts that do not touch the biosolids/soil mixture, feed crops, and fiber crops shall not be harvested for *30 days* after application of biosolids.

Animal Grazing

Animals shall not be grazed on the land for *30 days* after application of biosolids.

Turf Growing

Turf grown on land where biosolids are applied shall not be harvested for *1 year* after application of the biosolids when the harvested turf is placed on either land with a high potential for public exposure or a lawn, unless otherwise specified by the permitting authority.

Public Access

Public access to land with a high potential for public exposure shall be restricted for *1 year* after application of biosolids.

Public access to land with a low potential for public exposure shall be restricted for *30 days* after application of biosolids.

Processes to Significantly Reduce Pathogens Listed in Appendix B of 40 CFR Part 503

Aerobic Digestion

Biosolids arc agitated with air or oxygen to maintain aerobic conditions for a specific are mean cell residence time (MCRT) at a specific temperature. Values for the MCRT and temperature shall be between 40 days at 20°C and 60 days at 15°C.

Air Drying

Biosolids are dried on sand beds or on paved or unpaved basins. The biosolids dry for a minimum of 3 months. During 2 of the 3 months, the ambient average daily temperature is above 0°C.

Anaerobic Digestion

Biosolids are treated in the absence of air for a specific MCRT at a specific temperature. Values for the MCRT and temperature shall be between 15 days at 35°C to 55°C and 60 days at 20°C.

Composting

Using either the within-vessel, static-aerated pile, or windrow-composting methods, the temperature of the biosolids is raised to 40°C or higher and maintained for 5 days. For 4 hours during the 5-day period, the temperature in the compost pile exceeds 55°C.

Lime Stabilization

Sufficient lime is added to the biosolids to raise the pH of the biosolids to 12 after 2 hours of contact.

Summary of Biosolids Sampling Considerations

Factors to Consider in Developing a Sampling Program

Who must sample?

Preparer, land applier, surface disposer, or incinerator of biosolids

What must be sampled?

Biosolids:

- Metals (land application, surface disposal, incineration)
- Pathogens and vector attraction reduction (land application and surface disposal sites only)
- Nitrogen (land application only)

Biosolids incinerator emissions:

- Total hydrocarbons (or carbon monoxide), oxygen, temperature, information needed to determine moisture content, and mercury and beryllium, when applicable

Other:

- Methane gas in air (surface disposal sites only)

How often should sampling be done?

From once per year to once per month, depending on the amount of biosolids used or disposed.

How should sampling be done and how many samples should be taken?

Take either:

- Grab samples* (individual samples) for pathogens and percent volatile solids determinations, or
- Composite samples* (several grab samples combined) for metals.

No fixed number of individual samples required (except for class B pathogens, alternative 1, take seven samples). Enough material must be taken for the sample to be representative. Take a greater number of samples if there is a large amount of biosolids or if characteristics of biosolids vary significantly.

When to sample?

Before use or disposal. If biosolids are used or disposed before sampling results are available, and the results subsequently show that a regulatory limit is exceeded, the responsible person will be in noncompliance with USEPA Part 503 Biosolids Rule.

Where to collect samples?

Usually at site of preparer (e.g., treatment works). Sometimes samples must be collected at land application or surface disposal sites.

Sample from moving biosolids when possible to obtain a well-mixed sample. If you must sample from a stationary location, the sample should represent the entire area. Appropriate sampling points differ for liquid or dewatered biosolids (see the table titled Sampling Points for Biosolids (page 366).

What size of sample, sample equipment, storage times?

See the table titled Proper Conditions for Biosolids Sampling (page 367).

What methods should be used to analyze samples?

Part 503 requires that specific analytical methods be used for different types of samples.

---

* For guidance only; not a Part 503 rule requirement.

Sampling Points for Biosolids

Biosolids Type	Sampling Point
Anaerobically digested	Collect sample from taps on the discharge side of positive-displacement pumps.
Aerobically digested	Collect sample from taps on discharge lines from pumps. If batch digestion is used, collect sample directly from the digester. Cautions:
	1. If biosolids are aerated during sampling, air entrains in the sample. Volatile organic compounds may be purged with escaping air.
	2. When aeration is shut off, solids may settle rapidly.
Thickened	Collect sample from taps on the discharge side of positive-displacement pumps.
Heat treated	Collect sample from taps on the discharge side of positive-displacement pumps *after* decanting. Be careful when sampling heat-treated biosolids because of
	1. High tendency for solids separation
	2. High temperature of sample (temperature >60°C as sampled) can cause problems with certain sample containers due to cooling and subsequent contraction of entrained gases.
Dewatered, dried, composted, or thermally reduced	Collect sample from material collection conveyors and bulk containers. Collect sample from many locations within the biosolids mass and at various depths.
Dewatered by belt filter press, centrifuge, vacuum filter press	Collect sample from biosolids discharge chute.
Dewatered by biosolids press (plate and frame)	Collect sample from the storage bin; select four points within the storage bin, collect equal amount of sample from each point, and combine.
Dewatered by drying beds	Divide bed into quarters, grab equal amounts of sample from the center of each quarter, and combine to form a composite sample of the total bed. Each composite sample should include the entire depth of the biosolids material (down to the sand).
Compost piles	Collect sample directly from front-end loader while biosolids are being transported or stockpiled within a few days of use.

Proper Conditions for Biosolids Sampling

Parameter	Wide-Mouthed Container[*]	Preservative[†]	Maximum Storage Time[†]	Minimum Volume[‡]
Metals				
Solid and semisolid samples	P, G	Cool, 4°C	6 months	300 mL
Liquid (mercury only)	P, G	HNO_3 to pH <2	28 days	500 mL
Liquid (all other liquid metals)	P, G	HNO_3 to pH <2	6 months	1,000 mL
Pathogen Density and Vector Attraction Reduction				
Pathogens	G, P, B, SS	1. Cool in ice and water to <10°C if analysis delayed >1 hour, or	6 hours	1–4 L[§]
		2. Cool promptly to <4°C, or	24 hours (bacteria and viruses); 1 month (helminth ova)	
		3. Freeze and store samples to be analyzed for viruses at 0°C**	2 weeks	
Vector attraction reduction		Varies[‡]	Varies[‡]	1–4 L[§]

[*] P = plastic (polyethylene, polypropylene, polytetrafluoroethylene); G = glass (nonetched, heat-resistant); B = presterilized bags (for dewatered or free-flowing biosolids); SS = stainless steel (not steel- or zinc-coated).

[†] Preservatives should be added to sampling containers prior to actual sampling episodes. Storage times commence upon addition of sample to sampling container. Shipping of preserved samples to the laboratory may be, but is generally not, regulated under U.S. Department of Transportation hazardous materials regulations.

[‡] Varies with analytical method. Consult 40 CFR Parts 136 and 503.

[§] Reduced at the laboratory in ~300-mL samples.

** Do not freeze bacterial or helminth ova samples.

Discharge and Disinfection

The final discharge of wastewater requires compliance with regulations. Proper treatment is critical for the disinfection of wastewater and for environmental protection.

Reaction With Ammonia

Chloramines are formed in three successive steps, as shown in the following equations:

$$\underset{\text{(ammonia)}}{NH_3} \quad + \quad \underset{\substack{\text{(hypochlorous} \\ \text{acid)}}}{HOCl} \quad \rightarrow \quad \underset{\text{(monochloramine)}}{NH_2Cl} \quad + \quad H_2O$$

$$\underset{\text{(monochloramine)}}{NH_2Cl} \quad + \quad \underset{\substack{\text{(hypochlorous} \\ \text{acid)}}}{HOCl} \quad \rightarrow \quad \underset{\text{(dichloramine)}}{NHCl_2} \quad + \quad H_2O$$

$$\underset{\text{(dichloramine)}}{NHCl_2} \quad + \quad \underset{\substack{\text{(hypochlorous} \\ \text{acid)}}}{HOCl} \quad \rightarrow \quad \underset{\text{(trichloramine)}}{NCl_3} \quad + \quad H_2O$$

Reaction With Hydrogen Sulfide

When chlorine is used to remove hydrogen sulfide (H_2S), one of two reactions can occur, depending on the chlorine dosage:

$$\underset{\text{(chlorine)}}{Cl_2} \quad + \quad \underset{\text{(hydrogen sulfide)}}{H_2S} \quad \rightarrow \quad \underset{\text{(hydrochloric acid)}}{2HCl} \quad + \quad \underset{\text{(sulfur)}}{S}$$

or

$$\underset{\text{(chlorine)}}{4Cl_2} \ + \ \underset{\substack{\text{(hydrogen} \\ \text{sulfide)}}}{H_2S} \ + \ \underset{\text{(water)}}{4H_2O} \ \rightarrow \ \underset{\substack{\text{(hydrochloric} \\ \text{acid)}}}{8HCl} \ + \ \underset{\substack{\text{(sulfuric} \\ \text{acid)}}}{H_2SO_4}$$

Available Chlorine in NaOCl Solution

Percent Available Chlorine	Available Chlorine, *lb/gal*
10.0	0.833
12.5	1.04
15.0	1.25

When added to water, Ca(OCl)$_2$ reacts as follows:

Ca(OCl)$_2$	+	2H$_2$O	→	2HOCl	+	Ca(OH)$_2$
(calcium hypochlorite)		(water)		(hypochlorous acid)		(lime)

Sodium hypochlorite reacts with water to produce the desired HOCl as follows:

NaOCl	+	2H$_2$O	→	HOCl	+	NaOH
(sodium hypochlorite)		(water)		(hypochlorous acid)		(sodium hydroxide)

Residual

A basic equation for desired chlorine residual:

$$\begin{pmatrix} \text{chlorine dosage,} \\ \text{mg/L} \end{pmatrix} = \begin{pmatrix} \text{chlorine demand,} \\ \text{mg/L} \end{pmatrix} + \begin{pmatrix} \text{chlorine residual,} \\ \text{mg/L} \end{pmatrix}$$

Amounts of Chemicals Required to Obtain Various Chlorine Concentrations in 100,000 gal (378.5 m³) of Water*

Desired Chlorine Concentration in Water, mg/L	Chlorine Required		Sodium Hypochlorite Required						Calcium Hypochlorite Required	
			5% Available Chlorine		10% Available Chlorine		15% Available Chlorine		65% Available Chlorine	
	lb	(kg)	gal	(L)	gal	(L)	gal	(L)	lb	(kg)
2	1.7	(0.8)	3.9	(14.7)	2.0	(7.6)	1.3	(4.9)	2.6	(1.1)
10	8.3	(3.8)	19.4	(73.4)	9.9	(37.5)	6.7	(25.4)	12.8	(5.8)
50	42.0	(19.1)	97	(367.2)	49.6	(187.8)	33.4	(126.4)	64.0	(29.0)

* Amounts of sodium hypochlorite are based on concentrations of available chlorine by volume. For either sodium hypochlorite or calcium hypochlorite, extended or improper storage of chemicals may cause a loss of available chlorine.

Amounts of Chemicals Required to Obtain Various Chlorine Concentrations in 200 mg/L in Various Volumes of Water*

Volume of Water		Chlorine Required		Sodium Hypochlorite Required						Calcium Hypochlorite Required	
				5% Available Chlorine		10% Available Chlorine		15% Available Chlorine		65% Available Chlorine	
gal	(L)	lb	(kg)	gal	(L)	gal	(L)	gal	(L)	lb	(kg)
10	(37.9)	0.02	(9.1)	0.04	(0.15)	0.02	(0.08)	0.02	(0.08)	0.03	(13.6)
50	(189.3)	0.1	(45.4)	0.2	(0.76)	0.1	(0.38)	0.07	(0.26)	0.15	(68.0)
100	(378.5)	0.2	(90.7)	0.4	(1.51)	0.2	(0.76)	0.15	(0.57)	0.3	(136.1)
200	(757.1)	0.4	(181.4)	0.8	(3.03)	0.4	(1.51)	0.3	(1.14)	0.6	(272.2)

* Amounts of sodium hypochlorite are based on concentrations of available chlorine by volume. For either sodium hypochlorite or calcium hypochlorite, extended or improper storage of chemicals may cause a loss of available chlorine.

Number of 5-g Calcium Hypochlorite Tablets Required for Dose of 25 mg/L*

Pipe Diameter		Length of Pipe Section, *ft (m)*				
		≤13 (4.0)	18 (5.5)	20 (6.1)	30 (9.1)	40 (12.2)
in.	*(mm)*	Number of 5-g Calcium Hypochlorite Tablets				
4	(100)	1	1	1	1	1
6	(150)	1	1	1	2	2
8	(200)	1	2	2	3	4
10	(250)	2	3	3	4	5
12	(300)	3	4	4	6	7
16	(400)	4	6	7	10	13

*Based on 3.25-g available chlorine per tablet; any portion of tablet rounded to the next higher integer.

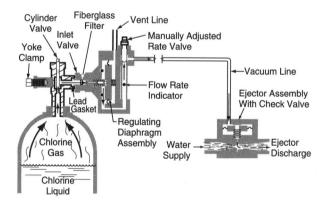

Gas Chlorinator

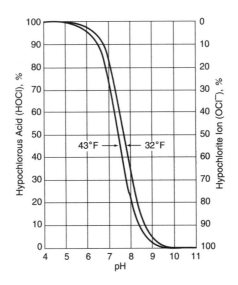

Relationship Among Hypochlorous Acid, Hypochlorite Ion, and pH

Chlorine Required to Produce 25-mg/L Concentration in 100 ft (30.5 m) of Pipe by Diameter

Pipe Diameter		100% Chlorine		1% Chlorine Solution	
in.	(mm)	lb	(g)	gal	(L)
4	(100)	.013	(5.9)	.16	(0.6)
6	(150)	.030	(13.6)	.36	(1.4)
8	(200)	.054	(24.5)	.65	(2.5)
10	(250)	.085	(38.6)	1.02	(3.9)
12	(300)	.120	(54.4)	1.44	(5.4)
16	(400)	.217	(98.4)	2.60	(9.8)

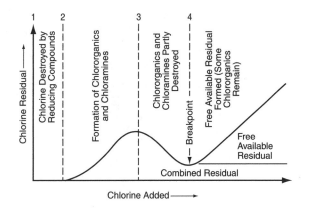

Breakpoint Chlorination Curve

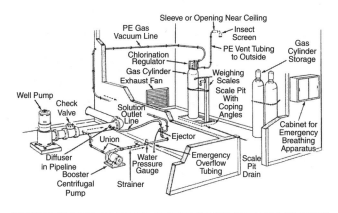

Typical Deep-Well Chlorination System

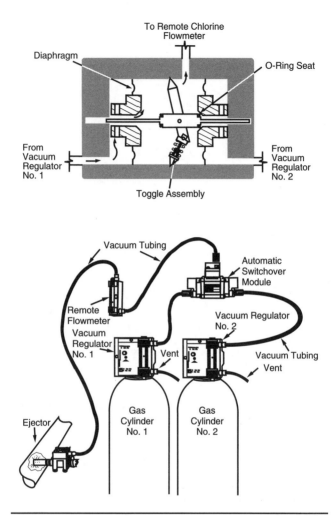

Typical Chlorinator Flow Diagrams

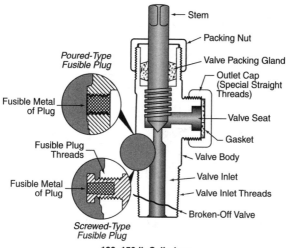

100–150-lb Cylinders

NOTE: Valve closes by turning clockwise; there are about 1¼ turns between wide-open and fully closed positions. All threads are right-hand threads.

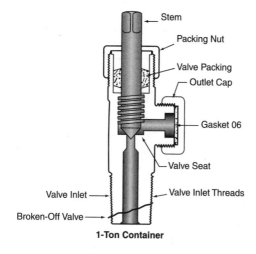

1-Ton Container

Standard Chlorine Cylinder Valves

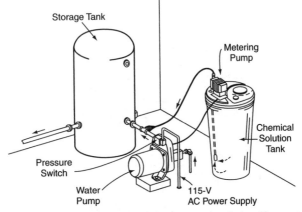

WARNING: When hazardous chemicals are pumped against positive pressure at point of application, use rigid pipe discharge line.

Courtesy of US Filter/Wallace & Tiernan.

Typical Hypochlorinator Installation

Summary of General Attributes of Chlorine Dioxide, Peracetic Acid, and UV Radiation

Attribute	ClO_2	Peracetic Acid	UV Radiation
Stability	Moderate	Low	High
Persistent residual	Moderate	None	None
Potential by-product formation	Yes	No	No
Reacts with ammonia	No	No	No
pH dependent	Moderate	No	No
Ease of operation	Moderate	Complex	Simple to complex
Temperature dependent	Moderate	Complex	Simple to complex
Contact time	Moderate	Low	Low
Safety concerns	High	High	Low
Effectiveness as bactericide	High	High	High
Effectiveness as viricide	High	High	High
Likelihood of regrowth	None	None	High

Recommended Chlorine Dosing Capacity for Various Types of Treatment Based on Design Average Flow

Type of Treatment	Illinois EPA Dosage, _mg/L_	GLUMRB* Dosage, _mg/L_
Primary settled effluent	20	
Lagoon effluent (unfiltered)	20	
Lagoon effluent (filtered)	10	
Trickling filter plant effluent	10	10
Activated sludge plant effluent	6	8
Activated sludge plant with chemical addition	4	
Nitrified effluent		6
Filtered effluent following mechanical biological treatment	4	6

*Great Lakes–Upper Mississippi River Board of State Public Health and Environmental Managers

Acute Values for Chlorine Toxicity in Aquatic Species

Species	Mean Acute Value, _g/L_
Freshwater	
Daphnia magna	27.66
Fathead minnow	105.2
Brook trout	117.4
Bluegill	245.8
Saltwater	
Menidia	37
Mysid	162

Discharge and Disinfection

Wastewater Characteristics Affecting Chlorination Performance

Wastewater Characteristic	Effects on Chlorine Disinfection
Ammonia	Forms chloramines when combined with chlorine
Biochemical oxygen demand	The degree of interference depends on their functional groups and chemical structures
Hardness, iron, nitrate	Minor effect, if any
Nitrite	Reduces effectiveness of chlorine and results in trihalomethanes
pH	Affects distribution between hypochlorous acid and hypochlorite ions and among the various chloramine species
Total suspended solids	Shielding of embedded bacteria and chlorine demand

ULTRAVIOLET LIGHT

Ultraviolet (UV) light rays cause death of microorganisms by oxidation of their enzymes. The most effective wavelength is 2,650 D (1 D = 10^{-8} cm). Thus, rays with a wavelength <3,100 D are effective. The mercury vapor lamp is an economical method of producing UV light of 2,537 D. For disinfection with UV light, water should be clear, colorless, and shallow (3–5 in. deep) to allow effective penetration of rays. These requirements as well as no residual effect and cost of application limit the use of this method to very small water supplies.

Some advantages and disadvantages of UV light are listed in the following sections.

Advantages

- UV disinfection is effective at inactivating most viruses, spores, and cysts.
- UV disinfection is a physical process rather than a chemical disinfectant, which eliminates the need to generate, handle, transport, or store toxic/hazardous or corrosive chemicals.
- There is no residual effect that can be harmful to humans or aquatic life.

- UV disinfection is user friendly for operators.
- UV disinfection has a shorter contact time when compared with other disinfectants (approximately 20–30 seconds with low-pressure lamps).
- UV disinfection equipment requires less space than other methods.

Disadvantages

- Low dosage may not effectively inactivate some viruses, spores, and cysts.
- Organisms can sometimes repair and reverse the destructive effects of UV through a "repair mechanism," known as photo reactivation, or in the absence of light known as "dark repair."
- A preventive maintenance program is necessary to control fouling of tubes.
- Turbidity and total suspended solids (TSS) in the wastewater can render UV disinfection ineffective. UV disinfection with low-pressure lamps is not as effective for secondary effluent with TSS levels above 30 mg/L.
- UV disinfection is not as cost-effective as chlorination, but costs are competitive when chlorination-dechlorination is used and fire codes are met.

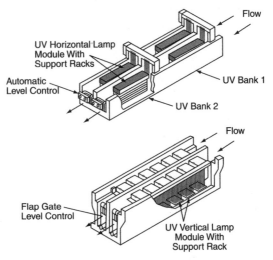

UV Horizontal Lamp Module With Support Racks

Automatic Level Control

Flow

UV Bank 1

UV Bank 2

Flow

Flap Gate Level Control

UV Vertical Lamp Module With Support Rack

NOTE: A UV bank is composed of a number of UV

Isometric Cut-Away Views of Typical UV Disinfection Systems

Wastewater Characteristics Affecting UV Disinfection Performance

Wastewater Characteristic	Effects on UV Disinfection
Ammonia	Minor effect, if any
Biochemical oxygen demand (BOD)	Minor effect, if any. If a large portion of the BOD is humic and/or unsaturated (or conjugated) compounds, however, then UV transmittance may be diminished.
Hardness	Affects solubility of metals that can absorb UV light. Can lead to the precipitation of carbonates on quartz tubes.
Humic materials, iron	High absorbency of UV radiation
Nitrate	Minor effect, if any
Nitrite	Minor effect, if any
pH	Affects solubility of metals and carbonates
Total suspended solids	Absorbs UV radiation and shields embedded bacteria

Mechanisms of Disinfection Using UV, Chlorine, and Ozone

UV	Chlorine	Ozone
1. Photochemical damage to RNA and DNA (e.g., formation of double bonds) within the cells of an organism.	1. Oxidation 2. Reactions with available chlorine 3. Protein precipitation	1. Direct oxidation/destruction of cell wall with leakage of cellular constituents outside of cell
2. The nucleic acids in microorganisms are the most important absorbers of the energy of light in the wavelength range of 240–280 nm.	4. Modification of cell wall permeability 5. Hydrolysis and mechanical disruption	2. Reactions with radical by-products of ozone decomposition 3. Damage to the constituents of the nucleic acids (purines and pyrimidines)
3. Because DNA and RNA carry genetic information for reproduction, damage of these substances can effectively inactivate the cell.		4. Breakage of carbon–nitrogen bonds leading to depolymerization

Discharge and Disinfection

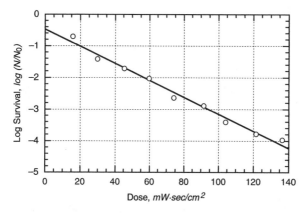

Copyright 1998 from *Wastewater Reclamation and Reuse*, edited by Takashi Asano. Reproduced by permission of Routledge/Taylor & Francis Group, LLC.

Typical Log Survival Versus Dose Curve of MS2 Coliphage Developed as Part of a Bioassay for the Measurement of UV Dose Within a UV Reactor

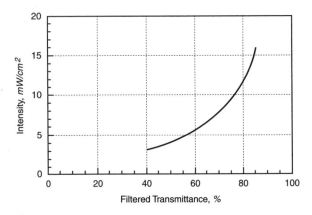

Average Intensity Within a 2-by-2 Lamp Array Calculated Using the Point Source Summation Method (based on a rate UV output of 26.7 W, 23-mm outside diameter quartz sleeve, and 75 mm center-line lamp spacing)

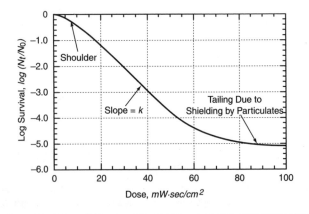

Log Survival Versus Dose Curve Illustrating the Tailing Region Generally Associated with Particulate Material in Wastewater

Comparison of Impact of Wastewater Characteristics on UV, Chlorine, and Ozone Disinfection

Wastewater Characteristic	UV Disinfection	Chlorine Disinfection	Ozone Disinfection
Ammonia	No or minor effect	Combines with chlorine to form chloramines	No or minor effect, can react at high pH
Biochemical oxygen demand (BOD), chemical oxygen demand (COD), etc.	No or minor effect, unless humic materials comprise a large portion of the BOD	Organic compounds that comprise the BOD and COD can exert a chlorine demand. The degree of interference depends on their functional groups and their chemical structure.	Organic compounds that comprise the BOD and COD can exert an ozone demand. The degree of interference depends on their functional groups and their chemical structure.
Hardness	Effects solubility of metals that may absorb UV light. Can lead to the precipitation of carbonates on quartz tubes.	No or minor effect	No or minor effect
Humic materials	Strong adsorbers of UV light	Reduces effectiveness of chlorine	Affects the rate of ozone decomposition and the ozone demand
Iron	Strong adsorber of UV light	No or minor effect	No or minor effect
Nitrate	No or minor effect	No or minor effect	Can reduce effectiveness of ozone
Nitrite	No or minor effect	Oxidized by chlorine	Oxidized by ozone
pH	Can effect solubility of metals and carbonates	Effects distribution between hypochlorous acid and hypochlorite ion	Effects the rate of ozone decomposition
Total suspended solids	UV absorption and shielding of embedded bacteria	Shielding of embedded bacteria	Increase ozone demand and shield embedded bacteria

Discharge and Disinfection

Typical Low-Intensity G64T5L UV Lamp Parameters and Performance Range

Parameter	Units	Standard Lamp Design	Performance Range
Nominal length[*]	Inches	64	NA[†]
Arc length	Inches	58	NA
Lamp wattage	Watts	65	NA
Lamp input current	Amperes	4.25×10^{-1}	3.0–5.25×10^{-1}
UV output per lamp at 253.7 nm[‡]	Watts	26.7	15.4–32.0
Lamp life: guaranteed (hours)[§]	Hours	8,760	4,000–13,000

[*] A 36-in. lamp is also available, although it is seldom used in designs.

[†] NA = not applicable.

[‡] Range of UV output is a function of lamp current and water temperature.

[§] Lamp manufacturer guarantees 8,760 hours; lamp life in field depends on lamp current and predetermined end of lamp life UV output intensity. The lower the operating current, the sooner the lamp will reach end of lamp life, typically defined as 65%–70% of new lamp intensity.

Suggested Rates of Wastewater Application

Soil Texture	Percolation rate, min/in. (min/cm)[*]	Application Rate, gpd/ft^2 (L/day/m^2)[†]
Gravel, coarse sand	<1 (<0.4)	Not suitable
Coarse to medium sand	1–5 (0.4–2.0)	1.2 (0.049)
Fine to loamy sand	6–15 (2.4–5.9)	0.8 (0.033)
Sandy loam to loam	16–30 (6.3–11.8)	0.6 (0.024)
Loam, porous silt	31–60 (12.2–23.6)	0.45 (0.018)
Silty clay loam, clay loam	61–120 (24.0–47.2)	0.2 (0.008)
Clay, colloidal clay	>120 (>47.2)	Not suitable

[*] min/in. $\times$ 0.4 = min/cm

[†] gpd/ft^2 $\times$ 40.8 = L/day/m^2

MARINE DISCHARGE

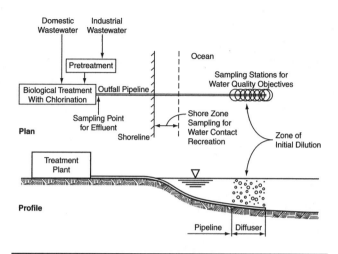

Schematic Plan and Profile Diagrams of Marine Discharge

Effluent Quality Limits of Major Wastewater Constituents for Ocean Discharge to Protect Marine Aquatic Life

Parameter	Monthly (30-day average)	Weekly (7-day average)	Maximum (at any time)
Grease and oil, *mg/L*	25	40	75
Suspended solids, *mg/L*	60 with a minimum removal of 75%		
Settleable solids, *mL/L*	1.0	1.5	3.0
Turbidity, *ntu*	75	100	225
pH	Within limits of 6.0–9.0 at all times		
Acute toxicity, *TUa**	1.5	2.0	2.5

$$*TUa = \frac{100}{96\text{-hour LC }50}$$

Where:

TUa = toxicity units acute

LC = lethal concentration 50%

Abbreviations
and Acronyms

In the wastewater industry as in other fields and disciplines, many names, titles, programs, organizations, legislative acts, measurements, and activities are abbreviated to reduce the volume of words and to simplify communications. In this section, common abbreviations and acronyms used in the water and wastewater industries—not only in this guide—are listed for easy reference.

Å	angstrom
A	ampere
AACE	American Association of Cost Engineers
AADF	annual average daily flow
AAS	atomic absorption spectrophotometry
AASHTO	American Association of State Highway and Transportation Officials
ABPA	American Backflow Prevention Association
ABS	alkylbenzene sulfonate; acrilonitrile butadiene styrene
AC	alternating current
A–C	asbestos cement
ACM	asbestos-containing material
acre-ft	acre-foot
ACS	American Chemical Society
ADA	Americans with Disabilities Act
ADF	average daily flow
AES	atomic emission spectroscopy
AHM	acutely hazardous material
A·hr	ampere-hour
AIChE	American Institute of Chemical Engineers
AIEE	American Institute of Electrical Engineers
AMWA	Association of Metropolitan Water Agencies
ANOVA	analysis of variance
ANPRM	Advanced Notice of Proposed Rulemaking
ANSI	American National Standards Institute
AOC	assimilable organic carbon
APHA	American Public Health Association
APLR	annual pollutant loading rate
APWA	American Public Works Association
ASCE	American Society of Civil Engineers
ASDWA	Association of State Drinking Water Administrators
ASME	American Society of Mechanical Engineers
ASSE	American Society of Safety Engineers; Association of State Sanitary Engineers
ASTM	American Society for Testing and Materials
atm	atmosphere
avdp or avoir.	avoirdupois
AWRA	American Water Resources Association
AWSAR	annual whole sludge (biosolids) application rate
AWWA	American Water Works Association
AwwaRF	Awwa Research Foundation

BAT	best available technology
bbl	barrel
BDOM	biodegradable organic matter
Bé	Baumé
BEAC	biologically enhanced activated carbon
BeV	billion electron volts
bgd	billion gallons per day
bhp	brake horsepower
bil gal	billion gallons
BNR	biological nutrient removal
BOD	biochemical oxygen demand *or* biological oxygen demand
BOM	background organic matter; biodegradable organic matter
bph	barrels per hour
bps	binary digits (bits) per second
Bq	becquerel (metric equivalent of curie)
BSA	bovine serum albumin
Btu	British thermal unit
bu	bushel
BV	bed volume
°C	degrees Celsius
C	coulomb
$C \times T$ *or* CT	disinfectant concentration $\times$ time
CAA	Clean Air Act
CBOD	carbonaceous biochemical oxygen demand
ccf	100 cubic feet
CCL	Contaminant Candidate List
CCR	consumer confidence report
cd	candela
CDC	Centers for Disease Control and Prevention
CERCLA	Comprehensive Environmental Response, Compensation, and Liability Act
CF	conventional filtration
cfm	cubic feet per minute
CFR	Code of Federal Regulations
cfs	cubic feet per second
cfu	colony-forming unit
CGPM	General Conference on Weights and Measures
Ci	curie
CI	cast iron

CIPP	cured in place product or pipe
C/kg	coulombs per kilogram
cm	centimeter
CMMS	computerized maintenance management system
COD	chemical oxygen demand
Co–Pt	chloroplatinate
CPLR	cumulative pollutant loading rate
cpm	counts per minute
CPP	concrete pressure pipe
cps	cycles per second (1 cps = 1 Hz)
CPSC	Consumer Products Safety Commission
cpu	chloroplatinate units
CPVC	chlorinated polyvinyl chloride
CSA	Canadian Standards Association
CT	contact time
CT or $C \times T$	disinfectant concentration $\times$ time
CTS-PE	copper tubing size polyethylene
cu	color unit; cubic
CUR	activated carbon usage rate
CWA	Clean Water Act
CWS	community water system

°	degree
d	day
D	dalton
da	darcy
DAF	dissolved air flotation
dB	decibel
DBCP	dibromochloropropane
DBP	disinfection by-product
DC	direct current
DCS	distributed control system
DCV	double check valve
DCVA	double check valve assembly
D/DBP	disinfectant/disinfection by-product
DDT	dichlorodiphenyltrichloroethane
DE	diatomaceous earth (filtration)
DI	ductile iron
diam.	diameter
DIPRA	Ductile Iron Pipe Research Association
dL	deciliter
DO	dissolved oxygen

DOC	dissolved organic carbon
DOT	Department of Transportation
DPD	N,N-diethyl-p-phenylenediamine
dr	dram
DSP	disodium phosphate
DTS	dry ton of solids
DWCCL	Drinking Water Contaminant Candidate List
DWV	drain, waste, and vent (pipe)
EBCT	empty-bed contact time
EC	electrical conductivity
ED	electrodialysis *or* effective diameter
EDB	ethylene dibromide
EDR	electrodialysis reversal
EDTA	ethylenediaminetetraacetic acid
EGL	energy grade line
EIS	Environmental Impact Statement
EJC	Engineers Joint Council
ElCD	electrolytic conductivity detector
emf	electromotive force
EPA	Environmental Protection Agency
EPCRA	Emergency Planning and Community Right-to-Know Act
EPDM	ethylene-propylene-diene-monomer
EPI-DMA	epichlorohydrin dimethylamine
EQ	exceptional quality
eq/L	equivalents per liter
ES	effective size
ESA	Endangered Species Act
ESWTR	Enhanced Surface Water Treatment Rule
eV	electron volt
°F	degrees Fahrenheit
F	farad
fbm	board feet (feet board measure)
FEMA	Federal Emergency Management Agency
FIFRA	Federal Insecticide, Fungicide, and Rodenticide Act
fl oz	fluid ounce
FM	Factory Mutual Engineering Corporation
F/M	food-to-microorganism ratio
fps	foot per second
FRP	fiberglass-reinforced plastic

ft	feet
ft/hr	feet per hour
ft/min	feet per minute
ft/sec	feet per second
ft/sec/ft	feet per second per foot
ft/sec^2	feet per second squared
ft^2/sec	feet squared per second
ft^2 *or* sq ft	square foot
ft^3/sec	cubic feet per second
ft^3 *or* cu ft	cubic feet
ft^3/hr *or* cu ft/hr	cubic feet per hour
ft^3/min *or* cu ft/min	cubic feet per minute
ft^3/sec *or* cu ft/sec	cubic feet per second
ft-lb	foot-pound
ftu	formazin turbidity unit
FY	fiscal year
g	gram
GAC	granular activated carbon
gal	gallon
gal/flush	gallons per flush
gal/ft^2	gallons per square foot
GAO	General Accounting office
GC	gas chromatography
GC–ECD	gas chromatography–electron capture detector
GC–MS	gas chromatography–mass spectrometry
GHT	garden hose thread
GIS	geographic information system
GL	gigaliter
GLUMRB	Great Lakes–Upper Mississippi River Board of State Public Health and Environmental Managers ("Ten States Standards")
gpcd	gallons per capita per day
gpd	gallons per day
gpd/ft^2	gallons per day per square foot
gpg	grains per gallon
gph	gallons per hour
gpm	gallons per minute
gpm/ft^2	gallons per minute per square foot
gps	gallons per second
GPS	global positioning system
gpy	gallons per year

gr	grain
GRP	glass-reinforced polyester
gsfd	gallons per square foot per day
GWUDI	groundwater under the direct influence of surface water
Gy	gray
H	henry
ha	hectare
HAA	haloacetic acid
HAA5	sum of five HAAs
HAN	haloacetonitrile
HAV	Hepatitis A virus
HDPE	high-density polyethylene
HDXLPE	high-density, cross-linked polyethylene
HF	hydrogen fluoride
HGL	hydraulic grade line
HGLE	hydraulic grade line elevation
HIV	human immunodeficiency virus
hL	hectoliter
hp	horsepower
HPC	heterotrophic plate count
hp·hr	horsepower-hour
HPLC	high-performance liquid chromatography
hr	hour
HRT	hydraulic retention time
HTH	High Test Hypochlorite
HVAC	heating, ventilating, and air conditioning
Hz	hertz
I&C	instrumentation and control
IBWA	International Bottled Water Association
ICP	inductively coupled plasma
ICR	Information Collection Rule
ID	inside diameter
IEEE	Institute of Electrical and Electronics Engineers
Imp	Imperial
in.	inch
in.-lb	inch-pound
in./min	inches per minute
in./sec	inches per second
in.^{2} *or* sq in.	square inch

in.³ *or* cu in.	cubic inches
IOC	inorganic contaminant
IP	iron pipe
IPS	iron pipe size
IPS-PE	iron pipe size polyethylene
IPT	iron pipe thread
IRC	International Research Center
ISA	Instrument Society of America
ISF	intermittent sand filter
ISO	International Organization for Standardization
IWRA	International Water Resources Association
J	joule
K	kelvin
kB	kilobyte
kg	kilogram
kHz	kilohertz (kilocycles)
kJ	kilojoule
km	kilometer
km²	square kilometers
kPa	kilopascal
kV	kilovolt
kVA	kilovolt-ampere
kvar	kiloreactive volt-ampere
kW	kilowatt
kW·hr	kilowatt-hour
L	liter
lb	pound
lb/day	pounds per day
lbf	pound force
lb/ft²	pounds per square foot
lbm	pound mass
LC	liquid chromatography
L/day	liters per day
LIN	liquid nitrogen
lin ft	linear feet
LLE	liquid–liquid extraction
lm	lumen
L/min	liters per minute
LOAEL	lowest-observed-adverse-effect level

LOX	liquid oxygen
LPG	liquefied petroleum gas
LSI	Langelier saturation index
LULU	locally unacceptable land use
lx	lux
m	meter
M	molar
m^2	square meters
m^3	cubic meters
mA	milliampere
mADC	milliampere direct current
max.	maximum
MB	megabyte
MBAS	methylene blue active substances
MCL	maximum contaminant level
MCLG	maximum contaminant level goal
MCRT	mean celled residence time
MDL	method detection limit
MDPE	medium-density polyethylene
meq	milliequivalent
meq/L	milliequivalents per liter
MeV	million electron volts
MF	membrane filter; microfiltration
MFL	million fibers per liter
mg	milligram
MG	million gallons
mgd	million gallons per day
mg/L	milligrams per liter
mhp	motor horsepower
MHz	megahertz (megacycles)
μ	micron
μg	microgram
μg/L	micrograms per liter
μm	micrometer
μM	micromolar
μmhos	micromhos
μmho/cm	micromhos per centimeter
μS	microsiemens
μW	microwatt
μW-sec/cm^2	microwatt-seconds per square centimeter
mi	mile

mi² *or* sq mi	square miles
mil	million
mil gal	million gallons
mμ	millimicron
min	minute
min.	minimum
mJ	millijoule
MJ	megajoule
MKS	meter/kilogram/second
mL	milliliter
ML	megaliter *or* million liters
MLSS	mixed liquor suspended solids
MLVSS	mixed liquor volatile suspended solids
mm	millimeter
m*M*	millimolar
mmol	millimole
mol	mole
mol wt	molecular weight
mol/L	moles per liter
MPC	maximum permissible concentration
mph	miles per hour
MPN	most probable number
mpy	mils per year
MRDL	maximum residual disinfectant level
MRDLG	maximum residual disinfectant level goal
mS	millisiemens
MS	mass spectrometry
MSDS	material safety data sheet
m/sec/m	meters per second per meter
MSL	mean sea level
MTD	maximally tolerated dose
MTF	multiple-tube fermentation
MTZ	mass transfer zone
MUD	municipal utility district
MW	molecular weight
MWCO	molecular weight cutoff
mW-sec	megawatts per second
N	newton
NA	not applicable; not analyzed
N/A	not applicable
NAS	National Academy of Science

NAWC	National Association of Water Companies
ND	not detected
NDWAC	National Drinking Water Advisory Council
NDWC	National Drinking Water Clearinghouse
NEC	National Electrical Code
NEMA	National Electrical Manufacturers Association
NEPA	National Environmental Policy Act
NESHAP	National Emission Standards for Hazardous Air Pollutants
NEWWA	New England Water Works Association
NF	nanofiltration
NFPA	National Fire Protection Association
ng/L	nanograms per liter
NGWA	National Ground Water Association
NH	American standard fire hose coupling thread (National hose thread)
NIOSH	National Institute of Occupational Safety and Health
NIPDWR	National Interim Primary Drinking Water Regulation
nm	nanometer
NOAEL	no-observed-adverse-effect level
NOM	natural organic matter
NPDES	National Pollutant Discharge Elimination System
NPDWR	National Primary Drinking Water Regulation
NPS	nominal pipe size; American standard straight pipe thread
NPSH	net positive suction head; American standard straight pipe for hose couplings (National pipe straight hose)
NPSHR	net positive suction head rate
NPSM	American standard straight pipe thread for free mechanical joints
NPT	American standard taper thread pipe (National pipe tapered)
NRWA	National Rural Water Association
NSDWR	National Secondary Drinking Water Regulation
NSFC	National Small Flows Clearinghouse
NST	American standard fire hose coupling thread (National standard thread)
NTNC	nontransient noncommunity
ntu	nephelometric turbidity unit
NWA	National Water Alliance
NWRA	National Water Resources Association

O&M	operations and maintenance
OD	outside diameter
ODM	maximum outside diameter
Ω	ohm
ORP	oxidation–reduction potential
OSHA	Occupational Safety and Health Administration
OUR	oxygen uptake rate
oz	ounce
ozf-in.	ounce-inch
Pa	pascal
P–A	presence–absence
PAA	peracetic acid
PAC	powdered activated carbon
PAH	polyaromatic hydrocarbon
Pa·sec	pascal-second
PB	polybutylene
PC	pollutant concentration
PCB	polychlorinated biphenyl
PCE	tetrachloroethylene (perchloroethylene)
pCi	picocurie
pCi/L	picocuries per liter
PCU	platinum–cobalt color unit
pE	oxidation–reduction (redox) potential
PE	polyethylene
PF	power factor
PFRP	process to further reduce pathogens
pfu	plaque-forming unit
pg	picogram
P&ID	process and instrumentation drawing
PID	proportional integral derivative control; photoionization detector
pk	peck
POE	point of entry
POTW	publicly owned treatment works
POU	point of use
PP	polypropylene
ppb	parts per billion
PPE	personal protective equipment
ppm	parts per million
ppt	parts per trillion; parts per thousand
PQL	practical quantitation level

PRI	primary
PRV	pressure-regulating valve
ps	picosecond
psi	pounds per square inch
psia	pounds per square inch absolute
psig	pounds per square inch gauge
PSRP	process to significantly reduce pathogens
Pt–Co	platinum–cobalt
PTFE	polytetrafluoroethylene
PVC	polyvinyl chloride
PVDF	polyvinylidene difluoride
PWD	public water district
PWL	pumping water level
PWS	public water system
QA	quality assessment
QC	quality control
qt	quart
r	roentgen
rad	radian
rad/sec	radians per second
RAS	return activated sludge
RBC	rotating biological contactor
RCRA	Resource Conservation and Recovery Act
RDL	reliable detection level
reg neg	regulatory negotiations
rem	roentgen equivalent, mammal
RMCL	recommended maximum contaminant level
RMP	risk management program
RO	reverse osmosis
rpm	revolutions per minute
rps	revolutions per second
RPZ	reduced pressure zone
RTD	resistance temperature detector
RTO	regenerative thermal oxidizer
RTP	reinforced thermoset plastic
RTU	remote terminal unit
S	siemens
SARA	Superfund Amendments and Reauthorization Act
SBR	sequencing batch reactor

SCADA	supervisory control and data acquisition
SCBA	self-contained breathing apparatus
SCD	streaming current detector
SCFM	standard cubic feet per minute
S/cm	siemens per centimeter
SDI	sludge density index
SDR	standard dimension ratio
SDWA	Safe Drinking Water Act
sec	second
sec^{-1}	inverse seconds
SEM	scanning electron microscope
SI	Système International d'Unités (International System of Units)
SMCL	secondary maximum contaminant level
SOC	synthetic organic chemical
SOUR	specific oxygen uptake rate
sp gr	specific gravity
sp ht	specific heat
SQL	Structured Query Language
sr	steradian
SSF	slow sand filtration
SUVA	specific ultraviolet absorbance
Sv	sievert
SVI	sludge volume index
SWD	side water depth
SWL	static water level
SWP	State Water Plan
SWTR	Surface Water Treatment Rule
t	metric ton *or* tonne
T	tesla
TC	thermocouple
TCE	trichloroethylene (or trichloroethene)
TCLP	toxic characteristic leaching procedure
TCR	Total Coliform Rule
tcu	true color unit
TDS	total dissolved solids
TF	trickling filter
TFE	tetrafluoroethylene
THM	trihalomethane
THMFP	trihalomethane formation potential
TNCWS	transient, noncommunity water system

TOC	total organic carbon
TON	threshold odor number; total organic nitrogen
TOX	total organic halogen
TPI	threads per inch
TS	total solids
TSCA	Toxic Substances Control Act
TSP	trisodium phosphate
TSPP	tetrasodium pyrophosphate
TSS	total suspended solids
TT	treatment technique
TTHM	total trihalomethanes
TVSS	transient voltage surge suppression
TWTDS	treatment works treating domestic sewage
uc	uniformity coefficient
UF	ultrafiltration
UFW	unaccounted-for water
UL	Underwriters Laboratories
UPS	uninterruptible power supply
URTH	unreasonable risk to health
USEPA	US Environmental Protection Agency
USPHS	US Public Health Service
UV	ultraviolet
V	volt
VA	volt-ampere
VAC	volts alternating current
VAR	volt-ampere-reactive; vector attraction reduction
VDC	volts direct current
VFD	variable-frequency drive
VOC	volatile organic compound
vol.	volume
VS	volatile solids
VSD	variable-speed drive
VSR	volatile solids reduction
W	watt
WAS	waste-activated sludge
Wb	weber
WEF	Water Environment Federation
WERL	Water Engineering Research Laboratory
WFP	Water For People

WHO	World Health Organization
whp	water horsepower
WHPA	wellhead protection area
WHPP	wellhead protection program
WIDB	Water Industry Data Base
WITAF	Water Industry Technical Action Fund
WQA	Water Quality Association
WQIC	Water Quality Information Center
wt	weight
WTP	water treatment plant
WWTP	wastewater treatment plant
Xe	xenon
yd	yard
yd^2 *or* sq yd	square yards
yd^3	cubic yards
ξp *or* zp	zeta potential

Glossary

From A *to* Z, *from* absolute pressure *to* zone of saturation *and everything in between, many terms used in the basic science—as well as the practical application of water and wastewater processes and technologies—are unique to the wastewater industry. For quick reference in the field, here is a compilation of wastewater quantity, quality, analysis, and useage terms, along with environmental and human-health-related terms commonly used in wastewater treatment.*

absolute pressure The total pressure in a system, including both the pressure of water and the pressure of the atmosphere (about 14.7 psi at sea level). Compare with *gauge pressure*.

acid Any substance that releases hydrogen ions (H^+) when it is mixed into water.

acidic solution A solution that contains significant numbers of H^+ ions.

aerobic Living or active in the presence of oxygen. Refers especially to microorganisms and/or decomposition of organic matter.

alkaline solution A solution that contains significant numbers of OH^- ions. A basic solution.

alkalinity A measurement of water's capacity to neutralize an acid. Compare *pH*.

ammeter An instrument for measuring amperes.

anaerobic Living or active in the absence of oxygen (e.g., anaerobic microorganisms).

animal (and poultry) manure Animal excreta, including bedding, feed, and other by-products of animal feeding and housing operations.

anion A negative ion.

annual average daily flow The average daily flow calculated using 1 year of data.

arithmetic mean A measurement of average value, calculated by summing all terms and dividing by the number of terms.

arithmetic scale A scale is a series of intervals (marks or lines), usually marked along the side and bottom of a graph, that represents the range of values of the data. When the marks or lines are equally spaced, it is called an arithmetic scale. Compare with *logarithmic scale*.

atom The smallest particle of an element that still retains the characteristics of that element.

atomic number The number of protons in the nucleus of an atom.

atomic weight The sum of the number of protons and the number of neutrons in the nucleus of an atom.

average daily flow A measurement of the amount of water treated by a plant each day. It is the average of the actual daily flows that occur within a period of time, such as a week, a month, or a year. Mathematically, it is the sum of all daily flows divided by the total number of daily flows used.

average flow rate The average of the instantaneous flow rates over a given period of time, such as a day.

bacteria Single-celled microscopic organisms lacking chlorophyll. Some cause disease and some do not. Some are involved in performing a variety of beneficial biological treatment processes including biological oxidation, solids digestion, nitrification, and denitrification.

balanced A chemical equation is balanced when, for each element in the equation, as many atoms are shown on the right side of the equation as are shown on the left side.

base   Any substance that releases hydroxyl ions (OH⁻) when it dissociates in water.

basic solution   A solution that contains significant numbers of OH⁻ ions.

battery   A device for producing DC electric current from a *chemical reaction*. In a storage battery, the process may be reversed, with current flowing into the battery, thus reversing the chemical reaction and recharging the battery.

bicarbonate alkalinity   Alkalinity caused by bicarbonate ions (HCO_3^-).

biochemical oxygen demand   The quantity of oxygen used in the biological and chemical oxidation (decomposition) of organic matter in a specified time, at a specified temperature (typically 5 days at 68°F [20°C]), and under specified conditions. A standardized biochemical oxygen demand test is used in assessing the amount of organic matter in wastewater.

biological oxidation   The aerobic degradation of organic substances by microorganisms, ultimately resulting in the production of carbon dioxide, water, microbial cells, and intermediate by-products.

biosolids   The organic solids product of municipal wastewater treatment that can be beneficially utilized. Wastewater treatment solids that have received processes to significantly reduce pathogens or processes to further reduce pathogens treatment, or their equivalents, according to the Part 503 rule to achieve a class A or class B pathogen status. The solids:liquid content of the product can vary: liquid biosolids, 1%–4% solids; thickened liquid biosolids, 4%–12% solids; dewatered biosolids, 12%–45% solids; dried biosolids, >50% solids (advanced alkaline stabilized, compost, thermally dried). In general, liquid biosolids and thickened liquids can be handled with a pump. Dewatered/dried biosolids are handled with a loader.

bond   See *chemical bond*.

brake horsepower   The power supplied to a pump by a motor. Compare with *water horsepower* and *motor horsepower*.

buffer   A substance capable in solution of resisting a reduction in pH as acid is added.

bulk density   The weight per standard volume (usually in pounds per cubic foot) of material as it would be shipped from the supplier to the treatment plant.

by-product   A secondary or additional product; something produced in the course of treating or manufacturing the principal product.

cake   Dewatered biosolids with a solids concentration high enough (>12%) to permit handling as a solid material. (NOTE: some dewatering agents might still cause slumping even with solids contents higher than 12%).

capacitance   A part of the total impedance of an electrical circuit tending to resist the flow of current. Capacitance can be added to cancel the effect of inductance. It is expressed in units of farads.

capacity   The flow rate that a pump is capable of producing.

carbonate alkalinity   Alkalinity caused by carbonate ions (CO_3^{-2}).

cation exchange capacity   A measure of the soil's capacity to attract and retain plant nutrients that occur in positively charged ionic form. Cation exchange capacity (CEC) is a focus of interest because fertilizers supply positively charged cationic plant nutrients, which are attracted to negatively charged anionic soil particles, including soil organic matter. Organically amended soils typically have a higher CEC (i.e., a higher capacity for attracting and retaining plant nutrients) than unamended or low organic soils.

cation   A positive ion.

chemical bond   The force that holds atoms together within molecules. A chemical bond is formed when a chemical reaction takes place. Two types of chemical bond are ionic bonds and covalent bonds.

chemical equation   A shorthand way, using chemical formulas, of writing the reaction that takes place when chemicals are brought together. The left side of the equation indicates the chemicals brought together (the reactants), the arrow indicates in which direction the reaction occurs, and the right side of the equation indicates the results (the products) of the chemical reaction.

chemical formula   See *formula*.

chemical reaction   A process that occurs when atoms of certain elements are brought together and combine to form molecules, or when molecules are broken down into individual atoms.

circuit breaker   A device that functions both as a current overload protective device and as a switch.

circumference   The distance measured around the outside edge of a circle.

composting   The accelerated decomposition of organic matter by microorganisms, which is accompanied by temperature increases above ambient; for biosolids, composting is typically a managed aerobic process.

compounds   Two or more elements bonded together by a chemical reaction.

concentration   In chemistry, a measurement of how much solute is contained in a given amount of solution (commonly measured in milligrams per liter).

conductor   A substance that permits the flow of electricity, especially one that conducts electricity with ease.

consolidated (biosolids)   A desirable characteristic of biosolids that allows the material to be stacked and remain nonflowing when stored.

converter   Generally, a DC generator driven by an AC motor.

covalent bond   A type of chemical bond in which electrons are shared. Compare with *ionic bond*.

critical control point   A location, event, or process point at which specific monitoring and responsive management practices should be applied.

cross multiplication   A method used to determine if two ratios are in proportion. In this method, the numerator of the first ratio is multiplied by the denominator of the second ratio. Similarly, the denominator of the first ratio is multiplied by the numerator of the second ratio. If the products of both multiplications are the same, the two ratios are in proportion to each other.

current regulator A device that automatically holds electric current within certain limits.

current The "flow rate" of electricity, measured in amperes. Compare with *potential*.

daily flow The volume of water that passes through a plant in 1 day (24 hours). More precisely called daily flow volume.

demand meter An instrument that measures the average power of a load during some specific interval.

denitrification The conversion of nitrogen compounds to nitrogen gas or nitrous oxide by microorganisms in the absence of oxygen.

denominator The part of a fraction below the line. A fraction indicates division of the numerator by the denominator.

density The weight of a substance per a unit of its volume (e.g., pounds per cubic foot or pounds per gallon).

design point The mark on the H–Q (head–capacity) curve of a pump characteristics curve that indicates the head and capacity at which the pump is intended to operate for best efficiency in a particular installation.

detention time The average length of time a drop of water or a suspended particle remains in a tank or chamber. Mathematically, it is the volume of water in the tank divided by the flow rate through the tank. The units of flow rate used in the calculation are dependent on whether the detention time is to be calculated in minutes, hours, or days.

dewatered biosolids The solid residue (12% total solids by weight or greater) remaining after removal of water from a liquid biosolids by draining, pressing, filtering, or centrifuging. Dewatering is distinguished from thickening in that dewatered biosolids may be transported by solids handling procedures.

diameter The length of a straight line measured through the center of a circle from one side to the other.

digestion The decomposition of organic matter by microorganisms with consequent volume reduction. Anaerobic digestion produces carbon dioxide and methane, whereas aerobic digestion produces carbon dioxide and water.

digit Any one of the 10 arabic numerals (0 through 9) by which all numbers may be expressed.

drawdown The amount the water level in a well drops once pumping begins. Drawdown equals static water level minus pumping water level.

dynamic discharge head The difference in height measured from the pump center line at the discharge of the pump to the point on the hydraulic grade line directly above it.

dynamic head See *total dynamic head*.

dynamic suction head The distance from the pump center line at the suction of the pump to the point of the hydraulic grade line directly above it.

Dynamic suction head exists only when the pump is below the piezometric surface of the water at the pump suction. When the pump is above the piezometric surface, the equivalent measurement is dynamic suction lift.

dynamic suction lift   The distance from the pump center line at the suction of the pump to the point on the hydraulic grade line directly below it. Dynamic suction lift exists only when the pump is above the piezometric surface of the water at the pump suction. When the pump is below the piezometric surface, the equivalent measurement is called dynamic suction head.

dynamic water system   The description of a water system when water is moving through the system.

effective height   The total feet of head against which a pump must work.

efficiency   The ratio of the total energy output to the total energy input, expressed as percent.

electromagnetics   The study of the combined effects of electricity and magnetism.

electron   One of the three elementary particles of an atom (along with protons and neutrons). An electron is a tiny, negatively charged particle that orbits around the nucleus of an atom. The number of electrons in the outermost shell is one of the most important characteristics of an atom in determining how chemically active an element will be and with what other elements or compounds it will react.

element   Any of more than 100 fundamental substances that consist of atoms of only one kind and that constitute all matter.

elevation head   The energy possessed per unit weight of a fluid because of its elevation above some reference point (called the reference datum). Elevation head is also called position head or potential head.

energy grade line   (Sometimes called energy gradient line or energy line.) A line joining the elevations of the energy heads; a line drawn above the hydraulic grade line by a distance equivalent to the velocity head of the flowing water at each section along a stream, channel, or conduit.

equivalent weight   The weight of an element or compound that, in a given chemical reaction, has the same combining capacity as 8 g of oxygen or as 1 g of hydrogen. The equivalent weight for an element or compound may vary with the reaction being considered.

eutrophication   A natural or artificial process of nutrient enrichment by which a water body becomes highly turbid, depleted in oxygen, and overgrown with undesirable algal blooms.

exceptional quality biosolids   Exceptional quality biosolids meet class A pathogen reduction; vector attraction reduction standards 1–8; and Part 503, Table 3, high-quality pollutant concentration standards.

exponent   An exponent indicates the number of times a base number is to be multiplied together. For example, a base number of 3 with an exponent of 5 is written 3^5. This indicates that the base number is to be multiplied together five times: $3^5 = 3 \times 3 \times 3 \times 3 \times 3$.

fecal coliform   The type of coliform bacteria present in virtually all fecal material produced by mammals. Since the fecal coliforms may not be pathogens, they indicate the potential presence of human disease organisms. See also *indicator organisms*.

fecal *Streptococcus*   A member of a group of gram-positive bacteria known as *Enterococci*, previously classified as a subgroup of *Streptococcus*. They are found in feces of humans, animals, and insects on plants often not in association with fecal contamination. See *indicator organisms*.

field storage   A temporary or seasonal storage area, usually located at the application site, which holds biosolids destined for use on designated fields. State regulations may or may not make distinctions between staging, stockpiling, or field storage. In addition, the time limits for the same material to be stored continuously on site before it must be land-applied range from 24 hours to 2 years.

filter backwash rate   A measurement of the volume of water flowing upward (backward) through a unit of filter surface area. Mathematically, it is the backwash flow rate divided by the total filter area.

filter loading rate   A measurement of the volume of water applied to each unit of filter surface area. Mathematically, it is the flow rate into the filter divided by the total filter area.

filter press   Equipment used near the end of the solids production process at a wastewater treatment facility to remove liquid from biosolids and produce a semisolid cake.

flow rate   A measure of the volume of water moving past a given point in a given period of time. Compare *instantaneous flow rate* and *average flow rate*.

formula weight   See *molecular weight*.

formula   Using the chemical symbols for each element, a formula is a shorthand way of writing what elements are present in a molecule and how many atoms of each element are present in each of the molecules. Also called a chemical formula.

friction head loss   The head lost by water flowing in a stream or conduit as the result of (1) the disturbance set up by the contact between the moving water and its containing conduit and (2) intermolecular friction.

fuse   A protective device that disconnects equipment from the power source when current exceeds a specified value.

gauge pressure   The water pressure as measured by a gauge. Gauge pressure is not the total pressure. Total water pressure (absolute pressure) also includes the atmospheric pressure (about 14.7 psi at sea level) exerted on the water. Gauge pressure in pounds per square inch is expressed as "psig."

generator   A piece of equipment used to transform rotary motion (for example, the output of a diesel engine) to electric current. Also, a person or organization who changes the biosolids characteristics either through treatment, mixing, or any other process.

good management practices Schedules of activities, operation and maintenance procedures (including practices to control odor, site runoff, spillage, leaks, or drainage), prohibitions, and other management practices found to be highly effective and practicable in the safe, community-friendly use of biosolids and in preventing or reducing discharge of pollutants to waters of the United States.

groups The vertical columns of elements in the periodic table.

head loss The amount of energy used by water in moving from one location to another.

head (1) A measure of the energy possessed by water at a given location in the water system, expressed in feet. (2) A measure of the pressure or force exerted by water, expressed in feet.

helminth and helminth ova Parasitic worms (e.g., roundworms, tapeworms, *Ascaris*, *Necator*, *Taenia*, and *Trichuris*) and ova (eggs) of these worms. Helminth ova are quite resistant to chlorination and can be passed out in the feces of infected humans and organisms and ingested with food or water. One helminth ovum is capable of hatching and growing when ingested.

homogeneous A term used to describe a substance with a uniform structure or composition throughout.

hydraulic loading rates Amount of water or liquid biosolids applied to a given treatment process and expressed as volume per unit time, or volume per unit time per surface area.

hydroxyl alkalinity Alkalinity caused by hydroxyl ions (OH^-).

indicator organisms Microorganisms, such as fecal coliforms and fecal streptococci (enterococci), used as surrogates for bacterial pathogens when testing biosolids, manure, compost, leachate, and water samples. Tests for the presence of the surrogates are used because they are relatively easy, rapid, and inexpensive compared to those required for pathogens such as *Salmonella* bacteria.

inductance An electrical property by which electrical energy is stored in a magnetic field. It is analogous to inertia in a hydraulic system. Induction has the effect of resisting changes in current flow. It is measured in henrys or meters squared kilograms per second squared per ampere squared.

infiltration The rate at which water enters the soil surface, expressed in inches per hour, influenced by both permeability and moisture content of the soil.

instantaneous flow rate A flow rate of water measured at one particular instant, such as by a metering device, involving the cross-sectional area of the channel or pipe and the velocity of the water at that instant.

insulator A substance that offers very great resistance, or hindrance, to the flow of electric current.

interpolation A technique used to determine values that fall between the marked intervals on a scale.

ion   An atom that is electrically unstable because it has more or fewer electrons than protons. A positive ion is called a cation. A negative ion is called an anion.

ionic bond   A type of chemical bond in which electrons are transferred. Compare with *covalent bond*.

isotopes   Atoms of the same element, but containing varying numbers of neutrons in the nucleus. For each element, the most common naturally occurring isotope is called the principal isotope of that element.

kill   The destruction of organisms in a water supply.

lagoon   A reservoir or pond built to contain water, sediment, and/or manure usually containing 4% to 12% solids until they can be removed for application to land.

land application   The spreading or spraying of biosolids onto the surface of land, the direct injection of biosolids below the soil surface, or the incorporation into the surface layer of soil. Also applies to manure and other organic residuals.

leachate   Liquid that has come into contact with or percolated through materials being stockpiled or stored; contains dissolved or suspended particles and nutrients.

liquid biosolids or manure   Biosolids or animal manure containing sufficient water (ordinarily more than 88%) to permit flow by gravity or pumping.

logarithmic scale (log scale)   A scale is a series of intervals (marks or lines), usually marked along the side and bottom of a graph, that represents the range of values of the data. When the marks or lines are varied logarithmically (and are therefore not equally spaced), the scale is called a logarithmic, or log, scale. Compare with *arithmetic scale*.

mercaptans   A group of volatile chemical compounds that are one of the breakdown products of sulfur-containing proteins. Noted for their disagreeable odor.

microorganism   Bacteria, fungi (molds, yeasts), protozoans, helminths, and viruses. The terms *microbe* and *microbial* are also used to refer to microorganisms, some of which cause disease, and others are beneficial. *Parasite* and *parasitic* refer to infectious protozoans and helminths. Microorganisms are ubiquitous, possess extremely high growth rates, and have the ability to degrade all naturally occurring organic compounds, including those in water and wastewater. They use organic matter for food.

mineralization   The process by which elements combined in organic form in living or dead organisms are eventually reconverted into inorganic forms to be made available for a new cycle of growth. The mineralization of organic compounds occurs through oxidation and metabolism by living microorganisms.

Glossary

minor head loss The energy losses that result from the resistance to flow as water passes through valves, fittings, inlets, and outlets of a piping system.

mitigation The act or state of reducing the severity, intensity, or harshness of something; to alleviate, diminish, or lessen, as to mitigate heat, cold, or odor.

mixture Two or more elements, compounds, or both, mixed together with no chemical reaction (bonding) occurring.

molality A measure of concentration defined as the number of moles of solute per liter of solvent. Not commonly used in wastewater treatment. Compare with *molarity*.

molarity A measure of concentration defined as the number of moles of solute per liter of solution.

molecular weight The sum of the atomic weights of all the atoms in the compound. Also called formula weight.

molecule Two or more atoms joined together by a chemical bond.

most probable number A statistical approximation of the number of microorganisms per unit volume or mass of sample. Often used to report the number of coliforms per 100 mL wastewater or water, but applicable to other microbial groups as well.

motor horsepower The horsepower equivalent to the watts of electric power supplied to a motor. Compare with *brake horsepower* and *water horsepower*.

municipal wastewater Household and commercial water discharged into municipal sewer pipes; contains mainly human excreta and used water. Distinguished from solely industrial wastewater.

neutralization The process of mixing an acid and a base to form a salt and water.

neutralize See *neutralization*.

neutron An uncharged elementary particle that has a mass approximately equal to that of the proton. Neutrons are present in all known atomic nuclei except the lightest hydrogen nucleus.

nitrification The biochemical oxidation of ammonia nitrogen to nitrate nitrogen, which is readily used by plants and microorganisms as a nutrient.

nomograph A graph in which three or more scales are used to solve mathematical problems.

nonpoint source pollution Human-made or human-induced alteration of the chemical, physical, biological, or radiological integrity of water or air, originating from any source other than a point source.

nonpoint source Any source, other than a point source, discharging pollutants into air or water.

normality A method of expressing the concentration of a solution. It is the number of equivalent weights of solute per liter of solution.

nucleus (plural: nuclei) The center of an atom, made up of positively charged particles called protons and uncharged particles called neutrons.

numerator The part of a fraction above the line. A fraction indicates division of the numerator by the denominator.

nutrient management plan A series of good management practices aimed at reducing agricultural nonpoint source pollution by balancing nutrient inputs with crop nutrient requirements. A plan includes soil testing; analysis of organic nutrient sources such as biosolids, compost, or animal manure; utilization of organic sources based on their nutrient content; estimation of realistic yield goals; nutrient recommendations based on soil test levels and yield goals; and optimal timing and method of nutrient applications.

nutrient Any substance that is assimilated by organisms and promotes growth; generally applied to nitrogen and phosphorus in wastewater but also other essential trace elements or organic compounds that microorganisms, plants, or animals use for their growth.

odor character The sensory quality of an odorant, defined by one or more descriptors, such as fecal (like manure), sweet, fishy, hay, woody resinous, musty, earthy.

odor dilutions to threshold or D/T A dimensionless unit expressing the strength of an odor. An odor requiring 500 binary (twofold) dilutions to reach the detection threshold has a D/T of 500. An odor with a D/T of 500 would be stronger than an odor with a D/T of 20.

odor intensity A measure of the perceived strength of an odor. This is determined by comparing the odorous sample with "standard" odors comprised of various concentrations of *n*-butanol in odor-free air.

odor pervasiveness Persistence of an odor; how noticeable an odorant is as its concentration changes; determined by serially diluting the odor and measuring intensity at each dilution.

odor threshold Detection—the minimum concentration of an odorant that, on average, can be detected in odor-free air. Recognition—the minimum concentration of an odorant that, on average, a person can distinguish by its definite character in a diluted sample.

off-site storage Storage of biosolids at locations away from the wastewater treatment plant or from the point of generation. Several terms encompass various types of storage: staging, stockpiling, field storage, and storage facility.

Ohm's law An equation expressing the relationship between the potential (E) in volts, the resistance (R) in ohms, and the current (I) in amperes for electricity passing through a metallic conductor. Ohm's law is $E = I \times R$.

organic compounds Generally, compounds containing carbon.

organics See *organic compounds*.

overland flow Refers to the free movement of water over the ground surface.

pathogen A disease-causing organism, including certain bacteria, fungi, helminths, protozoans, or viruses.

per capita Per person.

percent   The fraction of the whole expressed as parts per one hundred.

perimeter   The distance around the outer edge of a shape.

periodic table   A chart showing all elements arranged according to similarities of chemical properties.

periods   The horizontal rows of elements in a periodic table.

permeability   The rate of liquid movement through a unit cross section of saturated soil in unit time; commonly expressed in inches per hour.

pH   A measurement of how acidic or basic a substance is. The pH scale runs from 0 (most acidic) to 14 (most basic). The center of the range (7) indicates the substance is neutral, neither acidic or basic.

phytotoxin   Any substance having a toxic or poisonous effect on plant growth. Immature or anaerobic compost can contain volatile fatty acids that are phytotoxic to plants. Soluble salts can also be phytotoxic in addition to toxic heavy metals and toxic organic compounds.

point source   Any discernable, confined, or discrete conveyance from which pollutants are or may be discharged, including, but not limited to, any pipe, ditch, channel, tunnel, conduit, well, stack, container, rolling stock, concentrated animal feeding operation, or vessel or other floating craft.

pole   One end of a magnet (the north or south pole).

polymer   A compound composed of repeating subunits used to aid in flocculating suspended particulates in wastewater into large clusters. This flocculation aids solids removal and enhances the removal of water from biosolids during dewatering processes.

potential   The "pressure" of electricity, measured in volts. Compare with *current*.

power (in hydraulics or electricity)   The measure of the amount of work done in a given period of time. The rate of doing work. Measured in watts or horsepower.

power (in mathematics)   See *exponent*.

pressure head   A measurement of the amount of energy in water due to water pressure.

pressure   The force pushing on a unit area. Normally pressure can be measured in pascals, pounds per square inch, or feet of head.

principal isotopes   See *isotopes*.

process to further reduce pathogens   The process management protocol prescribed in USEPA Part 503 used to achieve class A biosolids in which pathogens are reduced to undetectable levels. Composting, advanced alkaline stabilization, chemical fixation, and drying or heat treatment are some of the processes that can be used to meet Part 503 requirements for class A.

process to significantly reduce pathogens   The process management protocol prescribed in USEPA Part 503 used to achieve class B biosolids in which pathogen numbers are significantly reduced but are still present. Additional restrictions on the use and placement of class B biosolids ensure a level of safety equivalent to class A. Aerobic and anaerobic digestion, air drying, and lime stabilization are types of processes used to meet the class B pathogen density limit of less than 2,000,000 fecal coliforms/gram dry weight of total solids.

products   The results of a chemical reaction. The products of a reaction are shown on the right side of a chemical equation.

proportion (proportionate)   When the relationship between two numbers in a ratio is the same as that between two other numbers in another ratio, the two ratios are said to be in proportion, or proportionate.

proton   One of the three elementary particles of an atom (along with neutrons and electrons). The proton is a positively charged particle located in the nucleus of an atom. The number of protons in the nucleus of an atom determines the atomic number of that element.

protozoa   Single-celled microorganisms, many species of which can infect humans and cause disease. The infective forms are passed as cysts or oocysts in the feces of humans and animals and accumulate in flocculated solids. They are quite resistant to disinfection processes, such as chlorination, that eliminate most bacteria but are susceptible to destruction by drying.

pump center line   An imaginary line through the center of a pump.

pump characteristics curve   A curve or curves showing the interrelation of speed, dynamic head, capacity, brake horsepower, and efficiency of a pump.

pumping water level   The water level measured when the pump is in operation.

radicals   Groups of elements chemically bonded together and acting like single atoms or ions in their ability to form other compounds.

radius   The distance from the center of a circle to its edge. One half of the diameter.

ratio   A relationship between two numbers. A ratio may be expressed using colons (for example, 1:2 or 3:7), or it may be expressed as a fraction (e.g., $\frac{1}{2}$ or $\frac{3}{7}$).

reactance   The combined effect of capacitance and inductance.

reactants   The chemicals brought together in a chemical reaction. The chemical reactants are shown on the left side of a chemical equation.

retention time   The period of time that wastewater or biosolids take to pass through a particular part of a treatment process, calculated by dividing the volume of processing unit by the volume of material flowing per unit time.

risk assessment   A quantitative measure of the probability of the occurrence of an adverse health or environmental effect. Involves a multistep process that includes hazard identification, exposure assessment, dose-response evaluation, and risk characterization. The latter combines this information so that risk is calculated: risk = hazard × exposure.

risk, potential Refers to a description of the pathways and considerations involved in the occurrence of an event (or series of events) that may result in an adverse health or environmental effect.

rule of continuity The rule states that the flow (Q) that enters a system must also be the flow that leaves the system. Mathematically, this rule is generally stated as $Q_1 = Q_2$ or (because $Q = AV$), $A_1V_1 = A_2V_2$.

runoff That part of the precipitation that runs off the surface of a drainage area when it is not absorbed by the soil.

safety factor The percentage above which a rated electrical device cannot be operated without damage or shortened life.

Salmonella Rod-shaped bacteria of the genus *Salmonella*, many of which are pathogenic, causing food poisoning, typhoid, and paratyphoid fever in human beings; or causing other infectious diseases in warm-blooded animals. Can cause allergic reactions in susceptible humans, and sickness, including severe diarrhea with discharge of blood.

salts Compounds resulting from acid–base mixtures.

scientific notation A method by which any number can be expressed as a number between 1 and 9 multiplied by a power of 10.

septage Domestic sewage (liquid and solids) removed from septic tanks, cesspools, portable toilets, and marine sanitation devices; not commercial or industrial wastewater.

sewage, domestic Residual liquids and solids from households conveyed in municipal wastewater sewers; distinguished from wastewater carried in dedicated industrial sewers.

side water depth The depth of water measured along a vertical interior wall.

slumping Failure of a stockpile to retain a consolidated shape, usually due to insufficient dewatering of the biosolids. Slumping may result in biosolids movement beyond the boundaries of a designated stockpile area or may create handling difficulties when the materials are scooped up and loaded into spreaders.

solids In water and wastewater treatment, any dissolved, suspended, or volatile substance contained in or removed from water or wastewater.

solute The substance dissolved in a solution. Compare with *solvent*.

solution A liquid containing a dissolved substance. The liquid alone is called the solvent, the dissolved substance is called the solute. Together they are called a solution.

solvent The liquid used to dissolve a substance. See *solution*.

specific capacity A measurement of the well yield per unit (usually per foot) of drawdown. Mathematically, it is the well yield divided by the drawdown.

specific gravity The ratio of the density of a substance to a standard density. For solids and liquids, the density is compared with the density of water (62.4 lb/ft^3). For gases the density is compared with the density of air (0.075 lb/ft^3).

stability   The characteristics of a material that contribute to its resistance to decomposition by microbes and to generation of odorous metabolites. The relevant characteristics include the degree of organic matter decomposition, nutrient, moisture, and salts content, pH, and temperature. For biosolids, compost, or animal manure, stability is a general term used to describe the quality of the material taking into account its origin, processing, and intended use.

staging   The concurrent delivery and application of biosolids, allowing for the transfer of biosolids from transport vehicles to land application equipment. Dewatered materials may be off-loaded from delivery vehicles to temporary stockpiles to facilitate the loading of spreading equipment.

standard solution   A solution with an accurately known concentration, used in the lab to determine the properties of unknown solutions.

static discharge head   The difference in height between the pump center line and the level of the discharge free water surface.

static suction head   The difference in elevation between the pump center line and the free water surface of the reservoir feeding the pump. In the measurement of static suction head, the piezometric surface of the water at the suction side of the pump is higher than the pump; otherwise, static suction lift is measured.

static suction lift   The difference in elevation between the pump center line of a pump and the free water surface of the liquid being pumped. In a static suction lift measurement, the piezometric surface of the water at the suction side of the pump is lower than the pump; otherwise, static suction head is measured.

stockpiling   The holding of biosolids at an active field site long enough to accumulate sufficient material to complete the field application.

storage facility   An area of land or constructed facilities committed to hold biosolids until the material may be land-applied at on- or off-site locations; may be used to store biosolids for up to 2 years. However, most are managed so that biosolids come and go on a shorter cycle based on weather conditions, crop rotations, and land availability, equipment availability, or to accumulate sufficient material for efficient spreading operations.

storage   Placement of class A or class B biosolids in designated locations (other than the wastewater treatment plant) until material is land applied; referred to as field storage. See also *off-site storage*.

surface overflow rate   A measurement of the amount of water leaving a sedimentation tank per unit of tank surface area. Mathematically, it is the flow rate from the tank divided by the tank surface area.

switch   A device to manually disconnect electrical equipment from the power source.

threshold odor   See *odor threshold*.

thrust block   A mass of concrete, cast in place between a fitting to be anchored against thrust and the undisturbed soil at the side or bottom of the pipe trench.

thrust A force resulting from water under pressure and in motion. Thrust pushes against fittings, valves, and hydrants and can cause couplings to leak or to pull apart entirely.

total alkalinity The combined effect of hydroxyl alkalinity (OH^-), carbonate alkalinity (CO_3^-), and bicarbonate alkalinity (HCO_3^-).

total dynamic head The difference in height between the hydraulic grade line (HGL) on the discharge side of the pump and the HGL on the suction side of the pump. This head is a measure of the total energy that a pump must impart to the water to move it from one point to another.

total static head The total height that the pump must lift the water when moving it from one point to another. The vertical distance from the suction free water surface to the discharge free water surface.

trihalomethanes Certain organic compounds, sometimes formed when water containing natural organics is chlorinated. Some trihalomethanes, in large enough concentrations, may be carcinogenic.

turbulence Irregular atmospheric motion especially characterized by up-and-down currents. Increasing turbulence results in dilution of odors.

valence electrons The electrons in the outermost electron shells. These electrons are one of the most important factors in determining which atoms will combine with other atoms.

valence One or more numbers assigned to each element, indicating the ability of the element to enter into chemical reactions with other elements.

vector attraction reduction A process for reducing the attractiveness of biosolids to vectors in order to reduce the potential for transmitting diseases from pathogens in biosolids.

vector An agent such as an insect, bird, or animal that is capable of transporting pathogens.

velocity head A measurement of the amount of energy in water due to its velocity, or motion.

virus A microscopic, nonfilterable biological unit, technically not living but capable of reproduction inside cells of other living organisms, including bacteria, protozoa, plants, and animals.

volatile compound A substance that vaporizes at ambient temperature. Above-average heat can increase the volatilization (vaporization) rate and amount of many volatile substances.

voltmeter An instrument for measuring volts.

wastewater treatment The processes commonly used to render water safe for discharge into a US waterway: (1) Preliminary treatment includes removal of screenings, grit, grease, and floating solids; (2) Primary treatment includes removal of readily settleable organic solids. Fifty to sixty percent suspended solids are typically removed along with 25%–40% biochemical oxygen demand (BOD); (3) Secondary treatment involves use of biological processes along with settling; 85%–90% of BOD and suspended solids are removed during secondary treatment;

(4) Tertiary treatment involves the use of additional biological, physical, or chemical processes to remove more of the remaining nutrients and suspended solids.

water hammer   The potentially damaging slam, bang, or shudder that occurs in a pipe when a sudden change in water velocity (usually as a result of too rapidly starting a pump or operating a valve) creates a great increase in water pressure.

water horsepower   The portion of the power delivered to a pump that is actually used to lift water. Compare with *brake horsepower* and *motor horsepower*.

wattmeter   An instrument for measuring real power in watts.

weir overflow rate   A measurement of the flow rate of water over each foot of weir in a sedimentation tank or circular clarifier. Mathematically, it is the flow rate over the weir divided by the total length of the weir.

whole numbers   Any of the natural numbers, such as 1, 2, 3, etc.; the negative of these numbers, such as –1, –2, –3, etc.; and zero. Also called integers or counting numbers.

wire-to-water efficiency   The ratio of the total power input (electric current expressed as motor horsepower) to a motor and pump assembly, to the total power output (water horsepower); expressed as a percent.

work   The operation of a force over a specific distance.

Index

NOTE: Abbreviations and acronyms are listed on pages 390–404, glossary terms on pages 406–421, and units of measure on pages 14–29, and accordingly are not cited individually in this index.

Index

Index

Index

Index

Index